Essentials of

Polymer Flooding Technique

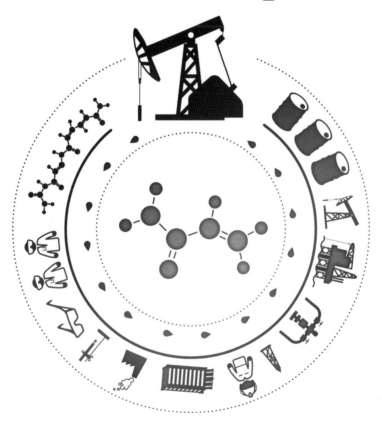

Antoine THOMAS

Essentials of Polymer Flooding Technique

Antoine Thomas

WILEY

This edition first published 2019
© 2019 John Wiley & Sons Ltd

The right of Antoine Thomas to be identified as the author of this work has been asserted in accordance with law.

Registered Offices
John Wiley & Sons, Inc., 111 River Street, Hoboken, NJ 07030, USA
John Wiley & Sons Ltd, The Atrium, Southern Gate, Chichester, West Sussex, PO19 8SQ, UK

Editorial Office
111 River Street, Hoboken, NJ 07030, USA

For details of our global editorial offices, customer services, and more information about Wiley products visit us at www.wiley.com.

Wiley also publishes its books in a variety of electronic formats and by print-on-demand. Some content that appears in standard print versions of this book may not be available in other formats.

Library of Congress Cataloging-in-Publication Data applied for

ISBN: 9781119537588

A catalogue record for this book is available from the British Library.

Cover Design: Wiley
Cover Images: Polymer Flooding illustration © Cyrille Cizel, Graphiste provided courtesy of Antoine Thomas

Set in 12/14.5pts TimesNewRomanMTStd by SPi Global, Chennai, India
Printed and bound in Singapore by Markono Print Media Pte Ltd

10 9 8 7 6 5 4 3 2 1

"…If you have an apple and I have an apple, and we swap apples — we each end up with only one apple. But if you and I have an idea and we swap ideas — we each end up with two ideas."

Charles F. Brannan

Table of Contents

Preface

Polymer flooding was first applied in the early 1960s. A spurt of applications of the process occurred between 1980 and 1986, but innovation was limited because those applications were dominated by tax considerations. However, beginning with the massive Daqing polymer flood in China in 1996, polymer flooding has experienced impressive innovation and growth in field applications. The author of this book works for a company (SNF) that was instrumental in most of the important field applications of polymer flooding throughout the world. As such, SNF acquired a unique perspective on the full range of topics associated with polymer flooding. That perspective is reflected in this book – especially in the last five chapters.

There are several key challenges whose solution would greatly aid the viability of polymer flooding. First, improvements are need in our ability to distribute the energy (induced pressure gradient) from a polymer drive deep into the reservoir (where the vast majority of oil resides). To date, this issue has largely been addressed by in-fill drilling – that is, placing injection and production wells closer together. Use of parallel horizontal wells has also been of value here. Even so, with existing polymer floods, we often must induce fractures in injection wells to allow economic injection rates for the viscous fluids. Polymer flooding could benefit greatly from improved characterization, placement, and exploitation of fractures (natural and induced) in reservoirs. This is especially true in less-permeable reservoirs.

A second major area for improvement is reducing retention (sometimes called adsorption) of polymers by the reservoir rock.

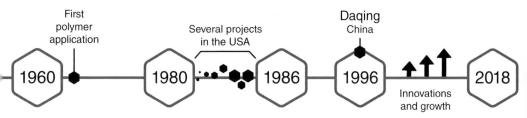

The polymer must penetrate deep into the porous rock of the reservoir in order to contact and displace the oil. If too much polymer is retained by the rock, the polymer may never penetrate sufficiently into the reservoir. Polymer retention can easily account for the largest economic hurdle in a polymer flood. In the past, laboratory studies (especially on outcrop rock) have often been overly optimistic about retention – especially in less-permeable rock and for associative polymers. Reduced polymer retention would be of great value.

A third important challenge is in expanding polymer flooding to hotter reservoirs. Great strides have been made in identifying monomers/polymers with sufficient stability for application in these reservoirs. However, the cost and viscosity associated with these polymers are often economically prohibitive. Improved manufacturing methods may be of substantial help here.

Treatment of produced polymer fluids is a fourth critical area for improvement. The viscous nature of polymer solutions often results in produced oil/water emulsions that are difficult to separate. Produced polymer has also been tied to other production problems. New methods to address these issues are needed. An ability to recycle produced polymers would also have value. Improved sampling of produced fluids is also needed, in that knowledge of whether the polymer propagates intact through a formation provides critical guidance to the operation and expansion of a polymer flood.

This book starts at a very basic level, for those with limited prior knowledge of petroleum production. The author's goal is to provide an easy-to-read introduction to the area of polymer flooding to improve oil production. The book also describes polymers to improve efficiency of chemical floods (involving surfactants and alkaline solutions). Chapters are short and end with a "nutshell" summary so the reader can quickly grasp the fundamentals. Each chapter also contains key references to allow more detailed examination of individual topics. The first few chapters provide brief introductions to oil recovery, chemical flooding methods, and polymer flooding. Chapter 4 lays out the important characteristics of polymers used for polymer flooding and important tests for their evaluation. Here, it is easy to overlook a crucial contribution that was made to polymer flooding technology by polymer manufacturers. In 1986, when oil prices collapsed from ~$30/bbl to ~$16/bbl, HPAM polymers typically cost about $2/lb. Most oil companies abandoned development of enhanced oil recovery processes because the chorus of oil company managers was, "Chemical flooding for oil recovery will never be viable because the price of polymers (and other chemicals) is tied to oil prices." However, because of innovations by polymer manufacturers, HPAM prices were commonly around $1/lb in 2012 when oil prices were over $100/bbl.

Chapters 5 through 9 provide a concise view of several key polymer-flooding topics that can't be found elsewhere. These are in the areas of pilot project design, field project engineering (water quality, oxygen removal, polymer dissolution equipment, filtration, pumps, and other equipment), produced water treatment, economics, and some important field case histories. Overall, this book is essential reading for anyone considering implementation of a polymer flood or chemical flood.

Randy Seright
January 2018

Abbreviations

«	Inches
$	Dollars
%	Percent
°C	Degrees Celsius
μ	Dynamic viscosity
μm	Micrometer
3D	Three dimensions
AIBN	Azobisisobutyronitrile
AMPS	Acrylamido-2-methylpropane sulfonic acid
API	American Petroleum Institute
AS	Alkali polymer injection
ASP	Alkali-surfactant-polymer injection
ATBS	Acrylamide tertiary butyl sulfonic acid
atm	Atmospheres
bbl	Barrels
BCF	Bioconcentration factor
bpd	Barrels per day
C_{30}	Molecule composed of 30 carbon atoms
CAPEX	Capital expenditures
CDG	Colloidal dispersion gel
CEOR	Chemical enhanced oil recovery
cm	Centimeters
cm^3	Cubic centimeters
cP	Centipoise
CSS	Cyclic steam simulation
D	Darcy
d	Days
Da	Daltons
DGF	Dissolved gas flotation
DOE	Department of Energy
DR	Drag reduction
EDTA	Ethylenediaminetetraacetic acid
EFSA	European food safety authority
Eh	Oxido-reduction potential

EOR	Enhanced oil recovery
ERDA	Energy Research and Development Administration
FCM	First-contact-miscible
Fe	Iron
FPSO	Floating production storage offloading
ft	Feet
ft/d	Feet/day
fw	Fractional flow
FWKO	Free-water knockout tank
g	Grams
g/L	Grams per liter
g/mol	Grams per mol
GPC	Gel permeation chromatography
H2S	Hydrogen sulfide
HOCNF	Harmonized Offshore Chemical Notification Format
HLB	Hydrophilic lipophilic balance
HPAM	Anionic polyacrylamide
HSE	Health safety and environment
IFT	Interfacial tension
IGF	Induced gas flotation
IOR	Improved oil recovery
k	Permeability
kcal	Kilocalories
kg	Kilograms
L	Liters
LCST	Lower critical solution temperature
m	Meters
$m.s^{-1}$	Meters per second
m^2/g	Square meters per gram
mD	Millidarcys
MF	Microfiltration
mL	Milliliters
mm	Millimeters
mN	Millinewtons
mPa	Millipascals
Mw	Molecular weight

N	Newtons
nm	Manometers
NEC	No effect concentration
NPV	Net present value
NVP	N Vinylpyrrolidone
O/W	Oil-in-water
OECD	Organization for Economic Co-operation and Development
OiW	Oil-in-water
OOIP	Oil originally in place
OPEX	Operational expenditures
P	Polymer injection
PAN	Polyacrylonitrile
PDI	Polydispersity index
pH	Potential of hydrogen
PLT	Production logging tool
ppb	Parts per billion
ppm	Parts per million
psi	Pounds per square inch
PSU	Polymer slicing unit
PV	Pore volume
PVDF	Polyvinylidene fluoride
Redox	Reduction/Oxidation
Rk	Residual resistance factor
Rm	Resistance factor
RO	Reverse osmosis
rpm	Rotations per minute
s^{-1}	Reciprocal second
SAC	Strong acid cation membrane
Sor	Residual oil saturation
SP	Surfactant polymer injection
SPE	Society of Petroleum Engineers
Sw	Water saturation
SWCTT	Single well chemical tracer test
Swi	Initial water saturation
TDS	Total dissolved salts
th. bbl	Thousand barrels

THPS	Tetrakis hydroxymethyl phosphonium sulphate
UF	Ultrafiltration
USA	United States of America
UV	Ultraviolet
VRR	Void replacement ratio
W	Watts
W/O	Water-in-oil
WAC	Weak acid cation membrane
WOR	Water-oil ratio
WSO	Water shut-off
wt	Weight
η	Kinematic viscosity

About the Author

Antoine THOMAS holds an MSc in petroleum geosciences from the Ecole Nationale Supérieure de Géologie in Nancy, France (2009). He joined SNF in 2011 as a reservoir engineer dealing with polymer flooding project design, implementation, and assistance for customers worldwide. In 2013, he spent part of his time in the R&D department, building the core flooding capacities for SNF and managing R&D projects in enhanced oil recovery (EOR) and hydraulic fracturing. He moved to Moscow in 2018 to supervise the oil and gas business from a technical standpoint, while maintaining contact with all SNF subsidiaries. He has published several papers and enjoys giving public lectures to share important learnings about EOR and hydraulic fracturing.

Thank you to all SNF reviewers and contributors who participated in the production of this book, including: Pascal Remy, René Pich, Nicolas Gaillard, Christophe Rivas, Julien Bonnier, Rémi Marchal, Flavien Gathier, Thierry Duteil, Dennis Marroni, Olivier Braun, Cédrick Favéro, Jean-Philippe Letullier … and the list continues. A special thank you to my North American reviewing team: Ryan Wilton, Kimberley McEwen, and Matthew Hopkins. Finally, a big thank you to Cyrille Cizel for putting everything together and creating the illustrations. Tremendous work.

Introduction

The energy spectrum of the world has changed dramatically over the last 100 years. Production and utilization of oil, the many offshoot industries it has spawned, and the technological advances developed have literally transformed the world as we see it today. The ubiquitous perception of abundant energy is also slowly changing, as the internet has brought information regarding the geopolitics of energy front and center.

However, have you ever asked people around you – your family, your friends, people at the fitness center – what percentage of oil can be extracted from a reservoir on average? Or, better yet, have you ever discussed with them their understanding of a geologic reservoir? You would probably be surprised to learn how many people think hydrocarbons can be recovered using a straw planted in a big, dark cavern full of oil or gas, or by shooting a bullet into the ground and having "black gold" bubble out. Moving from this fiction to reality requires education, science, time, and observation.

Moving hydrocarbons requires energy. The fossil fuels the world consumes on a daily basis are trapped in a porous material: an ancient, solid sponge formed by the accumulation of sediments over millions of years. What happens if you try to draw water from a sponge with a straw? It it slightly more difficult than simply pulling bulk fluid from a container. This same concept extends to hydrocarbon extraction.

Of the many available methods to produce hydrocarbon reserves, one involves water injection to sweep the oil toward producing wells. While widely deployed, this process (*waterflooding*) only helps recover approximately 35% of the oil contained in the giant "sponges."

35%! Really? That's not much.

With 65% of the resource stranded in place, engineers and scientists have worked for decades to develop technical solutions to recover it. Enhanced oil recovery (EOR) technologies have been implemented in various fields around the world, always using a case-by-case approach. One such technique consists of injecting viscosified water into the formation to displace the oil, instead of regular water. The viscosity contrast between the injected water and the viscous oil creates instability and promotes water penetration through the oil or complete bypass of the oil via geological highways (i.e. where the sponge or reservoir has the largest connected pores, making the flow much more easily). Increasing the viscosity of the water through the addition of water-soluble macromolecules (polymers) helps homogenize the displacement in the geologic formation: a larger volume of the sponge is contacted at the same time, leading to more efficient displacement and more oil being produced. This technique is called *polymer flooding*. It has been implemented since the late 1960s, with large commercial and technical success.

This book aims to summarize the key factors associated with polymers and polymer flooding – from the selection of the type of polymer through characterization techniques, to field design and implementation – discussing the main issues to consider when deploying this technology.

In an attempt to keep things simple, what follows is a pragmatic, rather than exhaustive, review of polymer flooding.

In terms of vocabulary, this is the last time you will read the word *sponge*; however, it is not the last time you will read the word *viscosity*!

Chapter one

Why Enhanced Oil Recovery?

In this chapter, the different production stages of an oil-bearing formation will be discussed with the goal of introducing enhanced oil recovery (EOR) techniques. Mainly, this chapter will discuss the common terminology used in the industry – which divides the life cycle of an asset into three stages (primary, secondary, and tertiary production) – to show the benefits of starting EOR techniques earlier in the development phase.

1.1. What Is a Reservoir?

The *reservoir* is an important component of a petroleum system. Oil and gas are formed from the decomposition of organic matter at high temperature and pressure in a source rock. Once formed, they can migrate upward until they either reach the surface and are degraded or are trapped by a seal or cap rock. If trapped, they tend to accumulate within a formation called a reservoir (Figure 1.1). Wells are drilled to reach this formation and start the extraction of the fluids.

A reservoir can be defined as subsurface rock formation having sufficient porosity and permeability to store and transmit fluids. Sedimentary rocks are the main formations of interest since they usually have higher porosity than magmatic and metamorphic rocks. Two categories are distinguished: clastic and carbonate rocks. *Clastic* rocks are formed from other existing rocks after erosion, transport, sedimentation, and burial. *Carbonate* rocks are mainly biogenic by origin: that is, they result from the accumulation of algae or microorganism remainders.

A good conventional reservoir is one with porosity and permeability high enough to allow the fluid to flow without much additional energy other than fluid expansion, reservoir compaction, or water injection.

Much attention has recently been directed toward so-called *unconventional reservoirs*, where it is necessary to adapt the technique to extract the hydrocarbons. This is the case for low permeability (tight) reservoirs or source rocks (shale gas and oil), where multi-stage, hydraulic fracturing is required to create paths to allow for more facile fluid drainage.

1.2. Hydrocarbon Recovery Mechanisms

Hydrocarbon production is commonly divided into three phases: primary, secondary, and tertiary (Figure 1.2).

Petroleum system and oil-bearing reservoirs

Figure 1.1

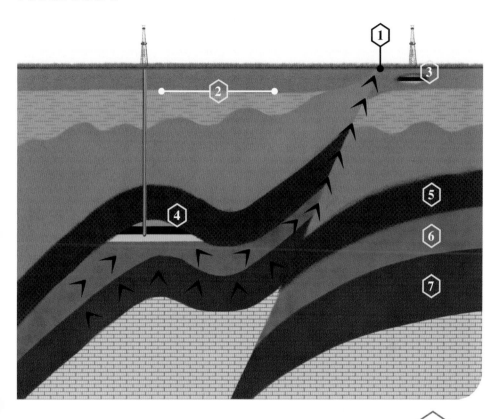

1. Oil seeps
2. Unconformity
3. Shallow reservoir
4. Reservoir

  Gas
  Oil
  Water

5. Seal rock
6. Reservoir rock
7. Source rock
> Migration

Figure 1.2 ## Hydrocarbon recovery mechanisms

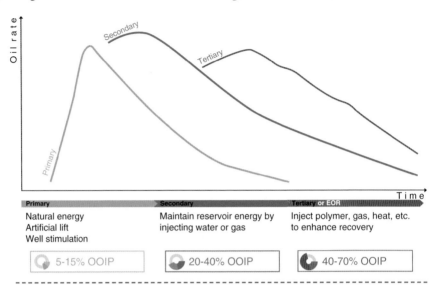

Primary recovery simply refers to the volume of hydrocarbons produced due to the natural energy prevailing in the reservoir or through artificial lift (i.e. pumping) through a single well. Common mechanisms behind primary recovery are as follows:

- Depletion drive

- Gas cap drive

- Gravity drainage

- Rock and/or liquid expansion

- Aquifer drive

The recovery factor at the end of this stage varies greatly depending upon reservoir and fluid characteristics. It can range from 5% to 40% or more in some cases. For heavy oil reservoirs or tight formations, the value is typically on the low end of this range.

Once the natural energy has been depleted, it is necessary to add energy to maintain or increase production levels to extract the remaining reserves. Thus, the secondary stage of recovery consists of introducing additional energy into the formation via one or several injection wells to drive or sweep the remaining fluids toward production wells. This secondary recovery process typically encompasses water or gas injections or the combination of both.

In the case of water injection, two main strategies may be implemented: (i) water injection for re-pressurizing and revitalizing the reservoir energy, and (ii) repeating pattern of injectors and producers forming a waterflood.

The tertiary or enhanced recovery stage of development can be significantly increased, reaching 50–60% for the most favorable reservoirs. However, with worldwide recovery factors averaging 35%, the study of techniques to enhance recovery of the remaining 65% left inside the formation is justified. For cases where new reservoir development is undertaken, secondary recovery could be implemented as enhanced oil recovery processes if waterflooding is forgone for transition directly to an EOR process. This could include, for example, a reservoir that is produced on primary production for a short period, after which polymer flood or cyclic steam injection is directly applied.

1.2.1. Anecdote

Between 1965 and 1979, there were five documented attempts to stimulate the production from hydrocarbon reservoirs by detonating nuclear devices in reservoir strata [1]. Three tests were performed in the United States and two in Russia, both aiming at increasing production rates and ultimate recovery from reservoirs. Subsurface explosive devices from 2.3 to 100 kt were used at depths from 1200 to 2560 m, creating post-shot problems: formation damage, radioactivity, creation of inflammable gases, and smaller-than-calculated fractured zones.

1.3. Definitions of IOR and EOR

Two acronyms are often encountered in the oil and gas industry when speaking about increasing the recovery of hydro-carbons: IOR for improved oil recovery and EOR for enhanced oil recovery [2]. IOR is a more general term, including any method toward increasing oil recovery (i.e. infill drilling, pressure support, operational and injection strategies, field redevelopment). EOR is usually considered a subset of IOR [3] and is often applied to reduce the oil saturation below the value obtained after waterflood, often referred to as the residual oil saturation (S_{or}) or, more specifically, residual oil saturation to waterflood (S_{orw}). Also, much interest has been focused on tertiary EOR. However, other definitions do not specifically tie this process to any specific production stage but rather include any method that can be used to increase the total recovery of any given field [4, 5].

1.4. What Controls Oil Recovery?

The efficiency of any recovery process can be defined by how much oil is contacted and displaced in a given reservoir (Figure 1.3). Recovery efficiency, E, is characterized as the product of two terms: (i) macroscopic sweep efficiency (mobilization at the reservoir scale, E_V) and (ii) microscopic sweep efficiency (mobilization at the pore scale, E_D – also known as the *displacement efficiency* [4]).

Macroscopic displacement efficiency relates to the volume of the reservoir contacted by the displacing fluid and is typically subdivided into areal and vertical macroscopic sweep efficiencies. This value is impacted by reservoir characteristics (geology, heterogeneities, fractures) and by fluid properties (viscosity, density). For example, it can be improved by maintaining a favorable mobility ratio between the displacing and displaced fluids by adding polymers to viscosify the injected water. This will be discussed in depth in subsequent chapters.

Areal and vertical sweep efficiency are parameters controlling oil recovery

Figure 1.3

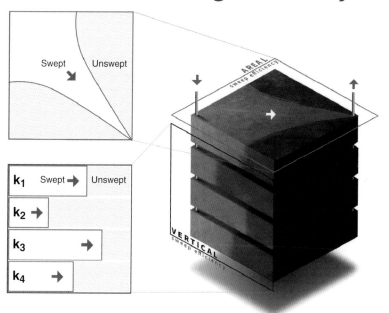

Microscopic displacement efficiency depends on the physical and chemical interactions that occur between the displacing fluid and oil. These include miscibility, wettability, and interfacial tension, which can be changed by adding specific additives to the injected fluid to dislodge the oil from the porous medium.

Equations (1.1) through (1.3) show the relationship and definition of all three efficiencies. E_D and E_V are typically expressed as fractions.

$$E = E_D E_V \qquad \dotfill (1.1)$$

$$E_D = \frac{\left(\dfrac{\overline{S_{oi}}}{B_{oi}} - \dfrac{\overline{S_o}}{B_o}\right)}{\dfrac{\overline{S_{oi}}}{B_{oi}}} = 1 - \left(\frac{S_{or}}{S_{oi}}\right)\left(\frac{B_{oi}}{B_o}\right) = 1 - \frac{S_{or}}{S_{oi}} \qquad \dotfill (1.2)$$

$$E_V = E_A E_I = \frac{N_{pwf}}{V_p \left(\dfrac{S_{oi}}{B_{oi}} - \dfrac{S_o}{B_o} \right)} \quad \dots\dots\dots\dots\dots\dots\dots\dots\dots\dots\dots (1.3)$$

where S_{oi}, S_o, and S_{or} are the initial oil saturation, oil saturation at time t, and residual oil saturation, respectively. Similarly, the initial and current oil formation volume factors, B_{oi} and B_o, represent the volume correction for expansion when fluids are brought to the surface. The terms V_p and N_{pwf} refer to the pore volume (void space containing fluids) and volume of oil recovered during waterflood, respectively.

It is obviously desirable for any EOR process that the values of E_D, E_V and, therefore, E, are maximized. From a practical standpoint, fluids that possess the ability to enhance both microscopic and macroscopic sweep efficiencies are difficult to develop. Many hurdles can be faced in developing and implementing such a fluid, including the following:

- Understanding of the reservoir's heterogeneities, geology, fractures, etc. Processes successfully designed in the laboratory can fail in the field because of geological factors and poor reservoir understanding.

- Flow in porous media and fluid interactions (mixing, shearing, adsorption of chemicals, etc.).

- Availability of the fluid or formulation, chemicals, etc. If the field considered is large, the volume of required chemicals can be *tremendous* and become an important limiting factor. Manufacturing, supply, logistics, and handling are the critical points to be assessed during the feasibility study, as these govern the actual delivery of chemicals to the remote site.

The choice of the most suitable EOR method requires an upfront clarification of expectations. Given the many uncertainties encountered throughout the process, it is illusory to expect a perfect EOR fluid formulation. De-risking can be achieved step-by-step through pilot tests and pragmatic approaches aimed at solving one problem at a time. A pilot project will be a critical step before considering full-field development.

1.5. Classification and Description of EOR Processes

EOR methods can be classified in several categories whose exact number depends on the authors and criteria. Green and Willhite [4] considered five categories (mobility-control, chemical, miscible, thermal, and other processes such as microbial EOR), while Lake [5] described three main subdivisions (thermal, chemical, and solvent methods). A good compromise would be four categories with thermal, chemical, miscible, and other EOR methods (microbial). Although this book is exclusively devoted to chemical methods and polymer flooding in particular, a brief description of each recovery technique will be given next.

1.5.1. Thermal Processes

Thermal processes include hot water injection, steam injection, and in situ combustion. Steam is used in two different ways: cyclic steam stimulation (CSS, Figure 1.4) or steam flood (Figure 1.5). Oil production is increased mainly due to thermal heat transfer, resulting in several mechanisms including oil viscosity reduction, oil swelling, and steam flashing.

For in situ combustion processes (Figure 1.6), air injection is implemented to generate thermal energy within the reservoir, and oil is recovered via viscosity reduction, fluid vaporization (light-end solvents, CO_2), or thermal cracking [6].

Figure 1.4 | Cyclic steam stimulation

1. Viscous oil

2. Heat zone

3. Condensed steam

4. Heated zone

5. Area heated by convection from hot water

6. Depleted oil sand

7. Condensed steam and thinned oil

8. Produced fluids

HUFF
(Injection phase)
days to weeks

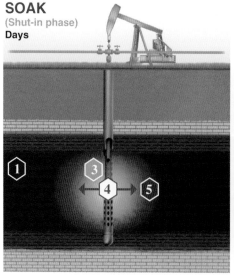

SOAK
(Shut-in phase)
Days

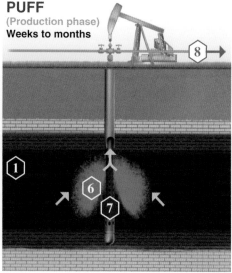

PUFF
(Production phase)
Weeks to months

Steam flood process

Figure 1.5

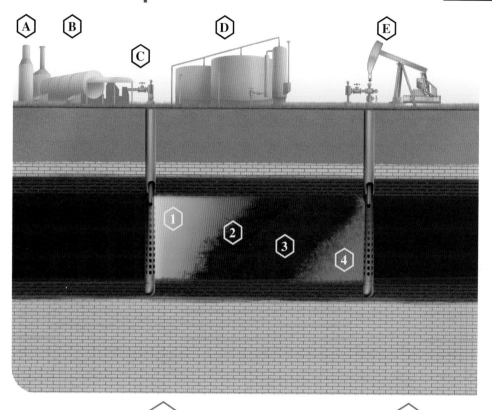

A. Stack gas scrubber

B. Steam generator

C. Injection well

D. Production fluids (oil, gas, and water) separation and storage facility

E. Production well

1. Steam and condensed water

2. Hot water

3. Oil bank

4. Oil and water zone near original reservoir temperature

Figure 1.6 In situ combustion process

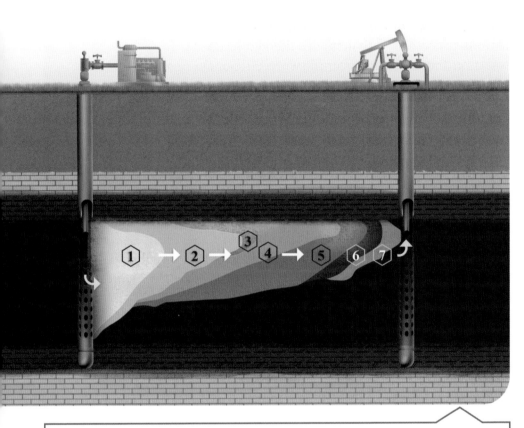

1. Injected air and water zone (burned out)

2. Air and vaporized water zone

3. Burning front and combustion zone (600°C–1200°F)

4. Steam or vaporizing zone (approx. 400°F)

5. Condensing or hot water zone (50°C– 200°F above initial temperature)

6. Oil bank (near initial temperature)

7. Cold combustion gases

1.5.2. Chemical Processes

This category includes the injection of water-soluble polymers, surfactants, and alkali alone or in combination, as well as other chemical cocktails such as microgels and nanogels, aimed at improving oil recovery from a given reservoir via conformance control. Polymers are used to viscosify the injection water and improve the overall sweep efficiency, E. Surfactants (surface active agents) are designed to lower the interfacial tension (IFT) between oil and water, mobilizing capillary-trapped oil. Alkali is used to synergistically improve the efficiency of surfactants via several mechanisms that will not be described in this section.

Microgels and nanogels are chemical technologies containing small polymer particles whose main goal is to decrease the permeability of thief zones, diverting the water to previously unswept areas. Their design and use is complex and requires a good reservoir understanding.

1.5.3. Miscible Processes

The objective here is to displace the oil with a fluid that is miscible in it, forming a single phase that can be moved through the reservoir (Figure 1.7). Green and Willhite [4] describe two categories:

- *First-contact-miscible (FCM)*. The injected fluid is directly miscible with the oil at the conditions of pressure and temperature encountered in the formation (i.e. liquefied petroleum gas);

- *Multiple-contact-miscible (MCM)*. The injected fluid is not miscible with the oil in the reservoir at first contact. Miscibility occurs when the proper conditions of pressure, temperature, and composition are reached (i.e. carbon dioxide); see Figure 1.7.

Figure 1.7 ## CO_2 miscible process

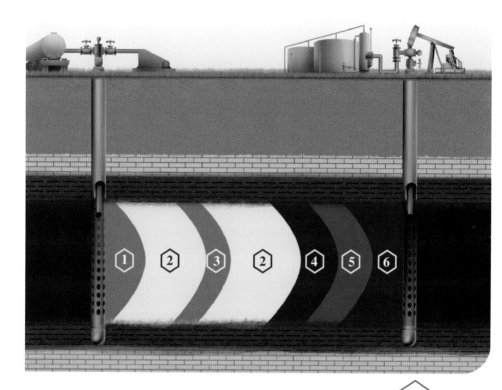

1. Drive water

2. CO_2

3. Water

4. Miscible zone

5. Oil bank

6. Additional oil recovery

Screening criteria for the applicability of each method will not be discussed here. References are given at the end of the chapter for further reading.

1.6. Why EOR? Cost, Reserve Replacement, and Recovery Factors

The budgets assigned to EOR developments often compete with other expenses, especially in terms of capital required for exploration (looking for undiscovered fields), new infill drills, injector conversions, maintenance programs (pump replacements, workovers) etc. However, several factors must be considered when evaluating EOR techniques against other development options [3]:

- Worldwide oil demand is forecast to increase in the long term.

- The discovery of new giant fields has greatly decreased as compared to past years, and current reserves are not being completely replaced.

- Drilling alone requires a large capital investment, and the drilling rate is not positively correlated with the discovery rate.

- The costs linked to exploration and extraction increase when targeting difficult reservoirs (ultra-deep offshore, arctic circle, other unconventional developments).

- Many EOR techniques have a long history, and uncertainty has decreased over time.

- More than 60% of oil remains after secondary recovery.

The benefit of any EOR technique is that it applies to every brownfield because:

- The resource has already been located, removing the requirement of further exploration.

- In many cases, the main infrastructure required is already in place.

- The markets for oil are available.

- It extends the life of the field and increases the size of the available resource.

As with greenfield developments, the objective is to maximize the final recovery from the outset of the development, knowing the current limits of secondary recovery techniques. Experience has shown that more than 60% of oil remains trapped in reservoirs after water injection. Why wait for production to become uneconomical before considering the so-called tertiary methods? Evaluations to collect the necessary data to model and appropriately de-risk the investment require time to implement and therefore should be started as early as possible. EOR techniques allow for accelerated oil production: the return on investment is quicker, which is beneficial when the time to exploit concessions is short.

Several techniques exist to improve recovery from oil reservoirs after primary and secondary production. Their applicability depends on many factors such as reservoir and fluid characteristics, field location, and accessibility. Until recently, EOR techniques were only considered late in the life of the field. This book will describe why their implementation should be studied much earlier in the development cycle and how they can provide a route to maximize recovery; proactive prevention is always better than a cure for an established predicament – in this case, an uneconomic and, ultimately shut-in petroleum field.

References

[1] Lorenz, J.C. (2001). The stimulation of hydrocarbon reservoirs with subsurface nuclear explosions. *Oil-Industry History* 2 (1).

[2] Stosur, G.J., Hite, R.J., Carnahan, N.F. et al. 2003. The alphabet soup of IOR, EOR and AOR: effective communication requires a definition of terms. Paper SPE84908 presented at the SPE International Improved Oil Recovery Conference in Asia Pacific, Kuala Lumpur, Malaysia, 20–21 October.

[3] Thomas, S. (2008). Enhanced oil recovery – an overview. *Oil & Gas Science and Technology* 63: 9–19.

[4] Willhite, G.P. and Green, D.W. (1998). *Enhanced Oil Recovery*. SPE Textbook Series, 6. Richardson, TX: Henry L. Doherty Memorial Fund of AIME, Society of Petroleum Engineers.

[5] Lake, L.W. (1989). *Enhanced Oil Recovery*, 323-324, 396-400. Englewood Cliffs, New Jersey: Prentice Hall.

[6] Donaldson, E.C., Chilingarian, G.V., and Yen, T.F. (1989). *Enhanced Oil Recovery, II – Processes and Operations*. Developments in Petroleum Science, vol. 17B, 604. Elsevier. ISBN: 0-444-42933-6.

Chapter two

Chemical Enhanced Oil Recovery Methods

In this chapter, the main chemical methods aimed at improving oil recovery will be described: polymer, surfactant-polymer, and alkali-surfactant-polymer flooding (P, SP, and ASP); microgels; and other conformance methods. A special focus will be placed on ASP flooding, with a brief knowledge summary including some field cases.

Essentials of Polymer Flooding Technique, First Edition. Antoine Thomas.
© 2019 John Wiley & Sons Ltd. Published 2019 by John Wiley & Sons Ltd.

2.1. Introduction

Chemical methods utilize a chemical formulation dissolved in fresh water or brine as the displacing fluid, which promotes a decrease in mobility ratio and/or an increase in the capillary number (definitions given in the following paragraphs). The change in mobility ratio is accomplished by adding water-soluble polyacrylamide to the injection water to increase its viscosity and improve the sweep efficiency inside the reservoir. Changing the capillary number is also possible by adding surface active agents (surfactants), which will decrease the interfacial tension and mobilize residual oil trapped in the reservoir. The addition of alkali promotes wettability changes, generates in situ surfactants (by saponification), and allows for decreasing surfactant concentration when considering ASP. These concepts will be discussed in more detail in this chapter.

Mobilization of oil is the key mechanism for any recovery process, with the saturation distribution of each fluid governed by both viscous and capillary forces. Figure 2.1 shows a schematic

Figure 2.1

▲ . a.Capillary schematic of water displacing oil

▲ . b.Capillary schematic of water pressure insufficient, residual oil saturation and capillary trapped oil droplet

(Adapted from Willhite and Green, 1998)

illustrating the concept of mobilizing and trapping residual oil saturation.

In this general two-liquid system of oil and water, oil would be considered the non-wetting phase, while water would be considered the wetting phase. Initially, oil is easy to displace from a capillary with low force and pressure requirements. In the second image (b) of the schematic in Figure 2.1, the water pressure is insufficient to overcome the capillary pressure within the trapped oil droplet, and residual oil saturation develops. The relationship between capillary pressure, interfacial tension (IFT), and tube radius is given by the capillary pressure equation, shown in its general form in Eq. (2.1).

Capillary pressure

$$P_c = \frac{2\sigma_{ow}\cos\theta}{r} \qquad\qquad (2.1)$$

The four main factors influencing oil phase capillary trapping are (i) changes in pore structure, i.e. radii; (ii) wettability alteration, i.e. fluid-rock interactions; (iii) fluid/fluid IFT; and (iv) fluid instabilities leading to viscous fingering and bypassing.

The correlation for a wetting fluid displacing a non-wetting fluid can be understood through the concept of capillary number, N_{ca} – which is best described by the ratio of viscous forces to capillary forces during the displacement process [1]. The capillary number can be shown in various forms: Eqs. (2.2) through (2.4) show these variations. The second form of the equation is the general form for a purely water-wet system where the wettability angle, $\theta = 0$; the third form introduces a viscosity ratio term to include the influence of oil viscosity.

Capillary number

$$N_{ca} = \frac{F_v}{F_c} = \frac{\mu * U}{\sigma_{ow}\cos\theta} \qquad\qquad (2.2)$$

$$N_{ca} = \frac{\mu_w * U}{\sigma_{ow}} \qquad\qquad (2.3)$$

$$N_{ca} = \frac{\mu_w * U}{\sigma_{ow} \cos\theta} \cdot \left(\frac{\mu_w}{\mu_o}\right)^{0.4} \quad\text{.. (2.4)}$$

where μ is the fluid viscosity (cP), U is the fluid's Darcy or superficial velocity (cm/sec), and σ is the surface or interfacial tension (dyne/cm). The subscripts denote oil (o) or water (w).

From the formula, it is intuitive that N_{ca} can be increased by increasing velocity, increasing water viscosity, or decreasing IFT between oil and water. In general, most flooding conditions operate such that a range between $10^{-6} < N_{ca} < 10^{-8}$, depending on whether the oil has high viscosity or detrimental IFT between water and oil liquids. In order to achieve this through velocity alone, unachievable field rates would be required to increase pore velocity sufficiently.

Depending on reservoir characteristics, field development, and economics, one or all three components in the ASP system can be considered and injected in the reservoir following specific sequences. Field implementation of alkaline without surfactants and polymers was initiated in 1925 [2]. Approximately 50 field trials were conducted prior to 1980 after it was discovered that surfactants could be generated in situ by interactions between alkaline chemicals and petroleum acids in the crude oil. The need for polymers to improve mobility control and sweep efficiency, and the addition of surfactants, were realized in the laboratory in the 1980s but implemented only in the 1990s.

2.2. Chemical EOR Methods

An overview of the main principles related to chemical enhanced oil recovery (EOR) techniques will be presented in the following paragraphs. Even though polymer flooding will be the main focus in subsequent chapters, all chemical EOR processes are introduced, as well as a comparison of polymer flooding with other chemical techniques such as crosslinked gels, colloidal dispersion gels (CDGs), and microgels.

2.2.1. Polymer Flooding

Polymer flooding consists of adding powder or emulsion polymers to the injection water of a waterflood, thereby viscosifying the water and decreasing its mobility. Even small reductions in aqueous phase mobility relative to the oil phase can mitigate the propensity for viscous instabilities. In addition to the increase in viscosity, some polymer can adhere to the reservoir rock and decrease the aqueous phase permeability, causing a lower mobility ratio. This results in greater vertical and areal reservoir sweep efficiency, which in turn translates into an accelerated (Figure 2.2) and often higher oil recovery factor at the end of the EOR process.

One of the key parameters to evaluate during screening of polymer floods is the mobility ratio, M. As the name implies, this

-----

Comparison of polymer flooding and waterflood: the injection of a viscous solution will accelerate oil production

Figure 2.2

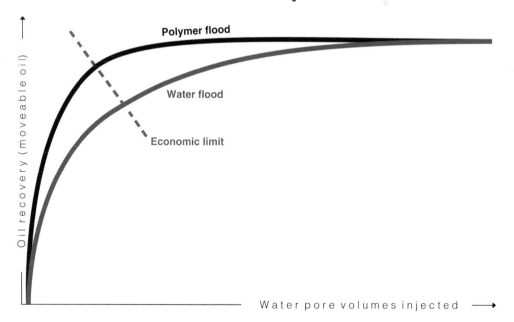

ratio comprises two mobilities: oil and water. In most discussions, *mobility ratio* usually refers to the end-point mobility ratio. Therefore, as shown in Eq. (2.5), the relative permeabilities of each fluid are used at their counterpart's irreducible saturation conditions, i.e. connate water saturation for oil and residual a.

Mobility ratio

$$M = \frac{\lambda_o}{\lambda_w} = \frac{\mu_o / k_{ro(Swc)}}{\mu_w / k_{rw(Sor)}} \qquad\qquad (2.5)$$

where λ, μ, and k_r are the mobility, viscosity, and relative end-point permeabilities, respectively, where the subscripts w and o refer to water and oil.

Polymer flooding can be applied as a tertiary recovery scheme, but more often than not, it is implemented after some period of secondary recovery under waterflood. Therefore, recovery under two main scenarios will occur:

- When the mobility ratio during a waterflood is not favorable (typically in viscous, heavy oils), continuous polymer injection can improve the sweep efficiency in the reservoir (greatly extending flood life).

- Even with a favorable mobility ratio, if the reservoir has some detrimental degree of heterogeneity in its stratigraphy, polymer injection can help reduce the water mobility in the high-permeability layers, supporting the displacement of oil from the low-permeability layers. This has implications in cases where layers are bounded by shales or are in vertical equilibrium and free crossflow exists.

For the first case, there is an inefficient macroscopic displacement that promotes early water breakthrough followed by a long period of two-phase production with increasing water-cut[1]. This situation can be illustrated by the concept of viscous fingering and ultimate bypassing of oil (Figure 2.3). This

Comparison of polymer flooding and waterflood: the injection of a viscous solution will accelerate oil production

Figure 2.3

or insufficiently viscous polymer are injected into heavy oil reservoirs, for instance.

It appears that even if the mobility ratio is favorable, the presence of high-permeability channels or large-scale reservoir layering and heterogeneities can greatly impair the areal and vertical sweep efficiencies during water injection. The presence of high-permeability layers will also lead to premature water breakthrough. In these reservoir cases, viscosification by polymer flooding can greatly improve the flooding conformance of the flood. For vertical injection, this may be more so in the vertical reservoir profile, as shown in Figure 2.4; while for horizontal injection, this may be better conformance along the wellbore length, as shown in Figure 2.5.

Figure 2.4 **Schematic illustration of vertical conformance improvement during vertical water and/or polymer injection under different recovery schemes**

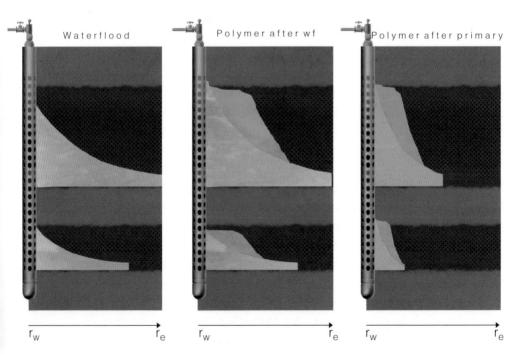

Schematic illustration of horizontal conformance improvement during horizontal water and/or polymer injection under tertiary or secondary recovery schemes

Figure 2.5

Polymer flooding
tertiary recovery

Polymer flooding
secondary recovery

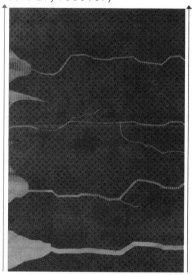
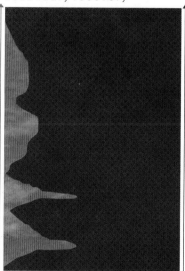

In either case, significant improvement in flooding efficiency can be gained by using polymers to increase the viscosity of the injected water. This aspect should be considered such that assets consisting of both heavy and light oil are not overlooked when screening reservoirs for polymer flooding candidates[3]. However, in all polymer flooding projects, breakthrough time of the polymer is not only related to the injected solution's viscosity; more important, the level of retention (polymer loss due to adsorption and entrapment mechanisms) and heterogeneity of the reservoir play a huge role in flood dynamics and recovery efficiency. This should also be carefully considered when determining EOR slug size (usually referred to as some percentage or fraction of overall area pore volume) to offset any chemical-rock interactions. This will be discussed in more detail in later chapters.

2.2.2. High-Viscosity Polymer Slugs

As of today, a great number of polymer injection EOR projects inject high-molecular-weight polyacrylamide-based polymer solutions with concentrations ranging between 1000 and 2000 ppm active. Depending on brine composition, this could yield injection wellhead viscosities as low as 5 to 50 cP and higher; therefore, polymer flooding technology has application to oils with a wide range of viscosities, presuming the reservoir permeability is such that it will accept larger-molecular-weight polymers. Decreasing the mobility ratio is an important aspect that can be improved through consideration of polymer chemistry, but reservoir heterogeneities can present a much larger challenge to recovery. Even if the mobility ratio appears to be favorable – for instance, in a light oil reservoir -- sweep efficiency can be greatly improved by increasing the target viscosity to balance the injection and displacement of oil between the low- and high-permeability layers. An example comes from the Daqing oil field in China, where polymer viscosities of 150–300 cP have been successfully injected into a reservoir containing an 11-cP crude oil, yielding oil recoveries above 20% oil originally in place (OOIP), similar to that reported for ASP injection [4, 5].

The main benefits of high-viscosity polymer slugs can be summarized as follows:

- A higher-viscosity polymer yields a better volumetric sweep efficiency, enlarging the zone contacted by the injected fluid. In heterogeneous systems with crossflow, the maximum viscosity above which no additional benefit is observed can be calculated using Darcy's law.

- A high-viscosity polymer acts like a "gel," decreasing the intake of higher-permeability zones.

- The higher the viscosity, the less the mixing with formation water or aquifer.

- A high-viscosity slug will help delay polymer breakthrough, given the increased swept area.

- Higher viscosity or higher polymer concentration helps compensate for possible polymer loss by retention.

Possible concerns about injectivity can be addressed by considering the injection strategy, including planning for lower injection/production rates or reducing well spacing.

In general, irreducible or residual oil saturation after polymer flooding cannot be reduced further beyond that possible with waterflooding. However, this aspect remains controversial; several authors have studied the effect of polymer viscoelasticity on oil recovery. This topic will be addressed later in this book. The greater and faster recovery constitutes the economic incentive for polymer flooding.

2.2.3. Surfactant-Polymer (SP)

Polymer and surfactants can be injected together to maximize oil recovery by taking advantage of macroscopic and microscopic sweep efficiencies. It is necessary to co-inject both chemicals to ensure that the polymer can conform to the chemical slug while the surfactant can release as much trapped oil as possible (Figure 2.6); this can be achieved only if a large reservoir zone is contacted by the cocktail. If surfactants are injected alone, there is a high risk that they may be prematurely lost to the reservoir through adsorption; in addition, depending on the viscosity of the chemical slug, they will follow the paths already created by water in the reservoir.

2.2.3.1. Surfactants

Surfactants are used to overcome immiscibility (i.e. the surface forces preventing single-phase mixing) between water and oil. *Surfactant* is a contraction of "surface active agent"; surfactants reduce the interfacial tension between oil and water, thereby stabilizing the mixture. Surfactants tend to spontaneously concentrate at the interface or "surface" between immiscible fluids, due to their chemical structure. A surfactant

Figure 2.6 Mobilization of trapped oil during a surfactant-polymer (SP) injection

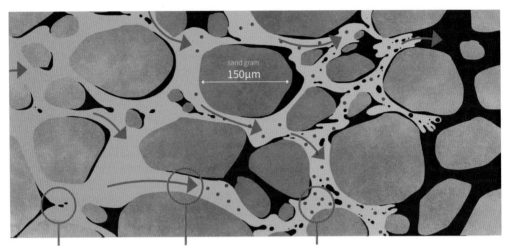

sand grain
150μm

Oil mobilization   Surfactant-polymer slug   Microemulsion

---

molecule will have a portion that is attracted to the oil phase (lyophilic or hydrophobic) and another portion that repels the oil phase and likes the water phase (lyophobic or hydrophilic) [6–9]. The affinity of these molecules is characterized by their hydrophilic-lipophilic balance (HLB) (Figure 2.7).

Surfactants are used for solubilization and to reduce interfacial tension. A lot of products currently exist, but efforts are ongoing to design cheaper, largely available, cost-effective

---

Figure 2.7 Hydrophilic lipophilic balance (HLB)

Solubilization

3,5   6 7 8   15   18   ⟶ HLB

W/O emulsifier   O/W emulsifier

$HLB = 20 \, X(M_h/M)$
M_h = molecular weight of hydrophilic groups
M = molecular weight of the molecule

molecules to make SP and ASP economical at a large scale. In SP processes, several surfactants and co-solvents are commonly mixed together to maximize oil recovery.

There are four main classes of surfactants: anionic, cationic, nonionic, and zwitterionic (or amphoteric) [10]. This classification depends upon the nature of the hydrophilic group (Figure 2.8).

Categories of surfactants used in chemical EOR

Figure 2.8

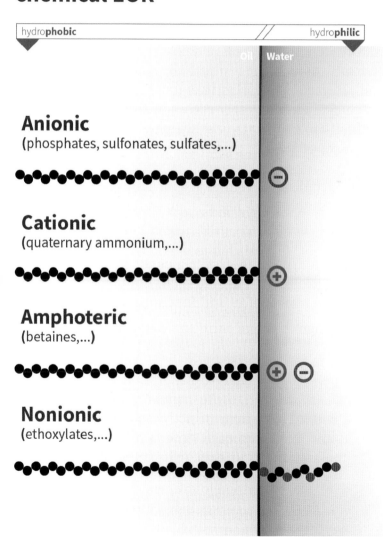

hydro**phobic**   hydro**philic**

Oil Water

Anionic
(phosphates, sulfonates, sulfates,...)

Cationic
(quaternary ammonium,...)

Amphoteric
(betaines,...)

Nonionic
(ethoxylates,...)

The first two are ionic species in aqueous solutions that carry, respectively, a negative or positive charge. The charge can be monovalent, as with sulfonates, or divalent with phosphates. Amphoteric surfactants can take on anionic or cationic characteristics depending upon the environment, such as the pH of the solvent.

Given the dual nature of surfactants, the charged species can not only interact with water and oil, but also interact with the mineralogy of the reservoir rock. Sand grains and clay-face surfaces are negatively charged. This can be problematic for cationic surfactants, as they are strongly attracted to negative charges and are, alone, not practical for chemical EOR: minimal surfactant would reside in the liquid phases to effectively reduce IFT sufficiently to mobilize residual oil.

Non-ionic surfactants are more tolerant to salinity and hardness as well as compatible with other classes of surfactants. A disadvantage of non-ionic surfactants is their physical form, with some being very viscous liquids or pastes. They are also generally more expensive than anionic surfactants, except for ethylene oxide derivatives, which display an inverse temperature effect on solubility (i.e. are less soluble at higher temperatures). They tend to have higher adsorption values than anionic surfactants.

Anionic surfactants have been most used in chemical EOR projects worldwide because of their commercial availability, cost, and amenable properties. This class contains a very wide range of molecular structures and functional groups including long-chain fatty acids, sulfosuccinates, alkyl sulfates, phosphates, and alkyl aryl sulfonates. Sulfonates are the most commonly used form for chemical EOR [11-14]. Anionic surfactants provide the greatest resistance to adsorption onto negatively charged solid surfaces like those contained within reservoir rocks, which makes them quite attractive to maximize residual oil contact and to render the EOR process more viable.

Several manufacturing processes are used to obtain these categories [14]:

- *Alkylation.* Production of $C_{14} - C_{30}$ alkylaryl compounds for use as hydrophobes that interact favorably with crude oil. When sulfonated, these long-chain alkylaryl compounds are used as primary surfactants.

- *Alkoxylation.* Reaction of propylene oxide and ethylene oxide with active alcohols to produce non-ionic surfactants and intermediates for further reaction.

- *Sulfation/Sulfonation.* Addition of a sulfate group to an alkoxylated molecule or a sulfonate group to an alkylaryl molecule.

Without polymer included in the surfactant slug, no mobility control exists: the surfactant would likely finger into the oil bank or be lost to existing channels, and the reservoir sweep would be very poor. In addition, a surfactant that is strongly adsorbed can cause a wettability change to more water-wet, ultimately causing the water-relative permeability to increase [15]. Therefore, these aspects must be counterbalanced by mitigating some surfactant loss and decreasing the aqueous mobility with polymer [16]. Furthermore, the polymer in both the surfactant (SP) slug and the follow-up polymer drive slug (to maintain mobility control of SP slug) helps mitigate the effects of reservoir permeability variation and improves the overall sweep efficiency in the reservoir.

Typical surfactant concentrations used in chemical EOR are 0.1–0.5% (active) for ASP formulations and up to 2.0% for SP [17–19]. The cost of the chemicals (with high dosages) and economics limit the surfactant concentration.

2.2.3.2. Field Cases

Since 2003, approximately 10 SP flooding pilot tests have been carried out in China [20]. These target reservoirs were characterized by high-permeability sandstones, conglomerates, and high temperature and high salinities. A few SP flooding pilot tests were conducted in Gudong Block 6 and Gudong Block 7 of the Shengli oil field with good results. An SP flooding pilot test was carried out in the Shengli oil field in September 2003, with an improved oil recovery of 12.1% after waterflooding. In August

2006, an expanded SP flood was carried out in the southeast part of Gudong Block 6. A petroleum sulfonate and a non-ionic surfactant were combined with a salt-resistant polymer and injected over 0.5 pore volume, yielding an extra 10.3% oil after water injection. Other tests were carried out in Gudong blocks 3 and 4, with incremental oil recovery of 4.5% over the injection period.

A single well chemical tracer test (SWCTT) is often employed to determine near-wellbore oil saturations at various points in a well's life. Alcohol and ester species are measured in an injection/production cycle, and changes in oil saturation and chemical retention can usually be determined. This can be beneficial to EOR operations in that it has been used to determine changes in oil saturation before and after exposure to a designed chemical/polymer system. A SWCTT was utilized in a 2014 offshore project in Abu-Dhabi by the company Total [21]. A carefully designed chemical formulation was injected into a carbonate formation to assess injectivity and the feasibility of chemical enhanced oil recovery (CEOR) implementation offshore.

A single-well chemical injection test was also carried out by the Hungarian company MOL in the Algyo oil field in 2013, south of Budapest [19], followed by a two-well pilot in 2016. The SP formulation consisted of 1.5% homemade surfactant (from MOL) and 0.1% of an ATBS-based copolymer able to sustain 100 °C.

2.2.4. Alkali-Surfactant-Polymer Flooding (ASP)

2.2.4.1. Theory

It was recognized in the United States in the 1920s that certain alkaline agents would react with naturally occurring acidic components in crude oil to generate surfactants in situ and therefore improve oil recovery. The minimum acid content required in the crude oil for the process to be effective is 0.3 mg KOH/g of oil, to generate sufficient surfactants or soaps in situ to lower the IFT between the oil and the aqueous phases below 10^{-3} dyne/cm (or mN/m) so that the residual oil saturation in the porous media may be reduced [22-26]. When alkali-polymer flooding (AP) is considered but the in situ generated soaps are not sufficient in the reservoir to account for retention [27], additional surfactants with concentrations up to 0.5% in weight can be added to the alkaline slug to further enhance the displacement process (ASP). For SP and ASP, polymer is required to improve mobility control and ensure that a large part of the reservoir is contacted by all chemical components in the EOR slug (Figure 2.9). The latter

Alkali-surfactant-polymer process

Figure 2.9

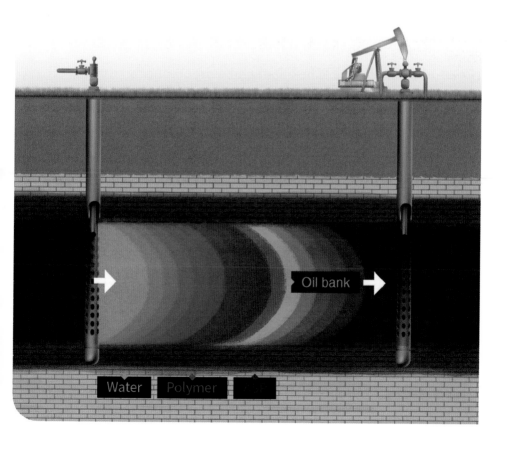

should be continued as long as economically feasible to prevent subsequent return to waterflood operations from fingering through the polymer slugs and compromising the overall flood of the upstream oil bank.

Commonly used alkaline agents are sodium hydroxide (NaOH), sodium carbonate (Na_2CO_3), ammonia, and sodium orthosilicate (Na_2O, SiO_2). The vast majority of field tests conducted prior to 2000 used NaOH because it is the most effective reagent when used with acidic crudes (this process is called *caustic flooding*). However, with caustic injection, the propensity for scaling problems increased due to brine incompatibilities at higher pH – these were found to cause severe problems in both the injection and production wells [28].

The successful application of an ASP project can theoretically recover between 10% and 30% incremental OOIP. Implementation of a chemical ASP flood is typically done in three steps, after the design and single-well tracer tests have been carried out [29]:

- *ASP stage.* The alkali and surfactant can mobilize oil, creating an oil bank that is "pushed" by the polymer. The latter also helps the formulation contact previously unswept zones in the reservoir. Typical slug sizes vary between 0.3 and 0.5 reservoir pore volumes.

- *Polymer stage.* Once the oil bank is created, the polymer chase continues to push it to the producer. Typical slug sizes also vary between 0.3 and 0.5 reservoir pore volume.

- *Water stage.* Strictly for economic purposes, the polymer stage is viscosity-tapered, eventually returning back to waterflood-only for the final stage, providing continued injection drive requirements until the end of the project.

2.2.4.2. Laboratory Studies

The laboratory studies required to design an ASP flood include several tasks that can be divided into static and dynamic types. This is a long and potentially costly process that includes studies such as the following:

- Acid analysis of the crude oil.
- Alkaline-oil interfacial visual tests.
- Alkaline-surfactant (AS) formulation tests with phase behavior studies and IFT measurements.
- Polymer stability and compatibility tests.
- ASP slug optimization-stability-clarity tests.
- Core flood experiments to quantify oil recovery and chemical retention.

The objective is to find a formulation that provides the largest single-phase emulsification of oil and water, indicating a low-IFT condition. A salinity scan can be used to find/obtain the phase behavior that may correspond to the best displacement of oil. Different emulsion types are obtained depending on composition, brine, and oil type, but in general the presence of a microemulsion (middle phase), also called a *Winsor III configuration*, is desired (see Figure 2.10). More details can be found in the references [22–25].

After this, 3D simulations can be created to clarify efficiency both from an economic standpoint and for reservoir optimization [24–26, 30].

A typical workflow is as follows:

a. *Reservoir conditions.* Analyze crude oil properties, rock, properties, temperature, oil, saturation, formation, and waterflood brine composition.

b. *Reservoir samples.* Collect crude oil and core samples from ASP target intervals.

c. *Polymer screening.* Select a polymer type and concentration. Tests: rheology, filtration ratio, screen factor, and temperature stability.

d. *Surfactant and alkali screening.* Select surfactant and alkali types and concentrations.

Figure 2.10 Looking for a Winsor III configuration. Salinity scan performed to find the optimum ASP design

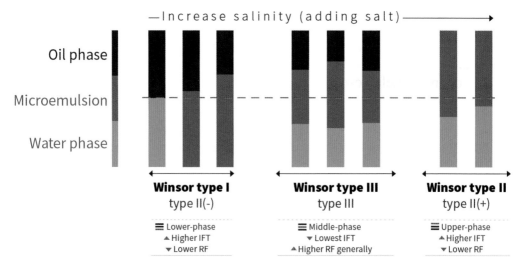

—Increase salinity (adding salt)—→

Oil phase

Microemulsion

Water phase

| Winsor type I | Winsor type III | Winsor type II |
| type II(-) | type III | type II(+) |

☰ Lower-phase	☰ Middle-phase	☰ Upper-phase
▲ Higher IFT	▼ Lowest IFT	▲ Higher IFT
▼ Lower RF	▲ Higher RF generally	▼ Lower RF

e. *Fluid/Fluid tests*. Solubility, temperature stability, phase behavior, ASP and micro emulsion rheology, adsorption, and consumption tests with crushed rock.

f. *Polymer coreflood or sandpack monophasic flow tests (no oil)*. Analyze polymer flow in the reservoir core. Tests: in situ rheology, polymer filtration properties, polymer adsorption, and resistance factor.

g. *Alkali/Surfactant coreflood or sandpack flow tests (no oil)*. Measure surfactant adsorption and alkali consumption in the field core. Tests: core flow and effluent analysis (titrations).

h. *ASP formulation*. Check compatibility of the ASP ingredients before core flow tests. Tests: solubility, phase separation, ASP phase behavior, emulsion viscosity, and core flow tests without oil.

i. *ASP coreflood or sandpack flow tests in a high-permeability outcrop core with dead crude*. Determine oil displacement efficiency in a high-permeability, low-clay-content outcrop rock (e.g. Berea, Bentheimer, or quartz sand).

j. *ASP coreflood or sandpack flow tests in an outcrop core with representative (reservoir) permeability.* Determine oil displacement efficiency in low-clay-content outcrop rock (e.g. Berea or Bentheimer) with representative permeability.

k. *ASP coreflood or sandpack flow tests in a field core with dead crude.* Measure oil displacement efficiency with dead crude in the field core.

l. *Confirm ASP effectiveness in phase tests and core flow tests with live crude.* Perform phase tests to determine optimal salinity with live crude. Measure oil displacement efficiency with live crude in the field core;

m. *Field tests.* Tracer test, single well test, pattern definition, etc.

n. *Multiple-well test.*

o. *Full field development.*

A typical timeframe for studies is as follows:

- Steps *a* through *e* : 6 months minimum
- Steps *f* through *l* : 12 months
- Step *m* : 1–2 months
- Step *n* : 2–5 years
- Step *o* : 10–20 years

From a field perspective, many other factors are considered important to the success of the process:

- *Distance.* The ASP process is very sensitive to well spacing, chemical retention, and injection rate, as these relate to the time to see a response and the size of slug required to achieve the maximum recovery factor. Most floods that do not achieve expectations could be due to insufficient lab evaluation/system design or could result from unknown geological factors beyond the distance reachable by tracers used in SWCTTs. The main concern is often the homogeneity and efficiency of the chemical slug beyond 50 m from the injection well.

- *Confinement and heterogeneity.* In many pilot tests, multi-well tracer tests are usually conducted to detect the confinement and heterogeneity in the reservoir. SWCTTs are performed to determine the residual oil saturation after the waterflood and also after chemical injection, as well as to infer chemical retention at the field scale.

- *Interpretation.* Field data taken from a SWCTT should be carefully interpreted before the field implementation. A successful SWCTT does not warrant successful field expansion, since a pilot is necessary to gather information for the actual oil rate, cumulative production, and, ultimately, potential recovery factor. Similarly, information on injectivity and consumption of chemicals and breakthrough events in the production wells can be obtained.

2.2.4.3. Economics

As of today, costs related to SP and ASP are relatively high given the chemical dosages injected and the extra work and equipment necessary to make the process viable. This is especially true for ASP processes where water softening is required to remove divalent ions to prevent scaling from occurring with the alkali. Today, the cost of surfactants is usually three to four times higher than polymer costs, with higher dosages yielding OPEX 5 to 15 times higher for SP than for polymer flooding and 2 to 7 times higher than for ASP. Other expenditures should also be considered, including these:

- Water softening for systems that use alkali.
- Chemical storage facilities for chemicals that are part of the formulation.
- Storage tanks for softened water.
- Waste storage tanks.

Obviously, an offshore ASP implementation would make the process even more complex due to water treatment; logistics; footprint; health, safety, and environment (HSE); and water disposal or reuse [31].

2.2.4.4. Field Cases

Worldwide, approximately 32 ASP field pilots and large-scale applications have been reported, with scarce performance data (Table 2.1). Among these 32 ASP projects, 12 projects were carried out in China, 6 in the United States, 8 in Canada (with important tax incentives), 3 in India, 1 in Oman, 1 in Russia, and 1 in Venezuela. India's Cairn reported an ASP project in Mangala [64] and Salym Petroleum Development in the West Salym oil field in Russia [66]. Several papers have been presented about the design and preparation of offshore ASP pilots in two fields: St. Joseph in Malaysia [67] and La Salina in Venezuela [59, 60, 68, 69].

Most of the field ASP projects were conducted in either five-spot patterns or inverted five-spot patterns, with an exception for heavy oil reservoirs. The Lagomar pilot was developed on an inverted seven-spot pattern; the Jilin Hong-Gang pilot was an inverted 13-spot pattern; and the Sa-bei pilot was a four-spot pattern. Most of the pilots were relatively small in scale, with few injectors. The largest ASP project so far was the Xingbei Xing-2-Zhong project in Daqing, with 17 injectors and 27 producers.

Almost all the ASP projects were carried out in sandstone reservoirs. For chemical EOR in carbonate reservoirs, many polymer projects were conducted between the 1960s and 1990s [70–72]. During this period, there were only a few SP projects, and no reported ASP projects. From the 1990s to 2000s, no chemical flood projects were reported, except four surfactant stimulation projects [73]. These projects are Mauddud in the Bahrain field [29], Cottonwood Creek in the Bighorn Basin of Wyoming [74, 75], the Yates field in Texas [71], and the Baturaja formation in the Semoga field in Indonesia [72]. A laboratory study was conducted to study the feasibility of ASP in the Cretaceous Upper

Table 2.1 **List of ASP floods, modified from ref.** [28]

Field	Country	References
Daqing Sa-zhong-xi (S-ZX)	China	[32–34]
Daqing Xing-wu- zhong (X5-Z)	China	[9, 32, 35]
Daqing Xing-2-xi (X2-X)	China	[36]
Daqing Sa-bei-1-xi (S-B)	China	[37]
Daqing Xing-bei xing-2-zhong (X2- Z)	China	[35, 38]
Daqing Sabei-bei-2-dong (SB-B2-D)	China	[35]
Shengli Gudong	China	[32, 39, 40]
Shengli Gudao-xi	China	[41–43]
Karamay	China	[43–46]
Jilin Hong-gang	China	[47]
Zhong-yuan Hu-zhuang-ji	China	[48]
Yumen Lao-jun- miao	China	[49]
Cambridge	USA	[50]
West Kiehl	USA	[51, 52]
Tanner	USA	[53]
Mellot Ranch	USA	[54]
Lawrence	USA	[26]
Sho-Vel-Tum	USA	
Mannville B	Canada	[55]
Little Bow	Canada	[56]
Gull Lake	Canada	Not available (NA)

Field	Country	References
Warner	Canada	NA
Instow	Canada	NA
Coleville	Canada	NA
Mooney	Canada	[57]
Suffield	Canada	[58]
Lagomar	Venezuela	[59–61]
Viraj	India	[62]
Jhalora	India	[63]
Mangala	India	[64]
Marmul	Oman	[65]
West Salym	Russia	[66]

Edwards reservoir, located in Central Texas, but no field trial was reported [70].

Canada's first field-scale ASP implementation was conducted at the Warner field, producing from the Mannville B Glauconitic sands in 2006 [55]. It was a follow-up project to an AP flood at a nearby field. The Warner reservoir has an average porosity of 25% and permeability above 1.5 D; the resident brine contains low total dissolved salts (TDS) of 5500 ppm, although due to divalents, softened water was used for all chemical injections. The Warner ASP injected 0.35 PV of an ASP slug consisting of 0.75 wt% sodium hydroxide (NaOH), 0.15 wt% anionic surfactant, and 0.12 wt% SNF Flopaam 3630 hydrolyzed polyacrylamide. After approximately 2.5 years of ASP injection, polymer-only taper was commenced in October 2008. Figure 2.10 shows the oil rate and oil cut observed through the waterflood infill program and includes the ASP response – first response was observed in less than a year (some wells in three months), but it was nearly two years before a peak oil rate of $300 \, \text{m}^3 \text{d}^{-1}$ (~1800 bopd) was achieved. The subsequent loss in oil rate was most

Figure 2.11 ASP production response at Warner Mannville B Glauconitic sand [55]

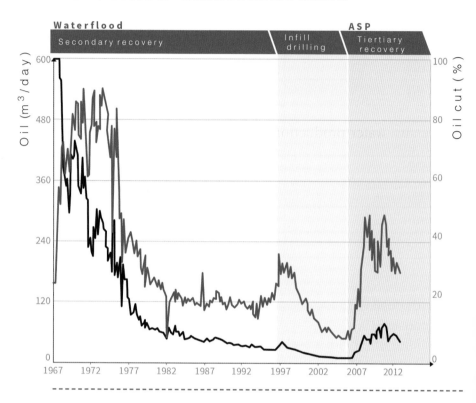

likely due to injectivity decline, lag in well responses not occurring all at the same time, and problematic operational issues related to silicate scale affecting producer run-time [11]; a follow-up paper provides more detail into the scaling issues and solutions [11]. A brief regain in oil rate coincided with a peak oil cut of ~11% in 2011, five years after ASP injection started and three years after switching to polymer-only injection (Figure 2.11).

Even though significant amounts of oil have been recovered in these applications, the amount of true incremental oil recovered by ASP flooding over the waterflood remains questionable, because many of these applications were implemented in the early stages of waterflooding and polymers were also used for the improvement of mobility control. To logically evaluate the incremental oil recovery from the ASP, a baseline of the ultimate waterflood and polymer flood recoveries must be established first.

2.2.5. Other Chemical Methods

2.2.5.1. Gels vs. Polymer Injection

Excessive water production in oil field operations is a major economic problem for operators and is inevitable once water-flooding or water pressure support recovery schemes are initiated. In some light oil fields, depending on the age of the waterflood, the viscosity of the oil, and the initial oil saturation conditions, water-to-oil ratios (WORs) may range from 0.1 to 5, representing a water-cut of 5–80%. For heavier oils in high-permeability, unconsolidated media, periods of low water-cut are typically short-lived, and much of the oil production can occur at WOR above 10 and great than 90% water-cut.

Two distinct types of water production exist. This first type of water production corresponds to the water that is co-produced with oil as part of the oil's fractional flow characteristics: the percentage usually increases during the exploitation of the field, especially as different layers break through and relative permeability and capillary forces begin to dominate production profiles. In good, clean sands that are homogeneous and contain low-viscosity, high American Petroleum Institute (API) oils, water breakthrough can be prolonged.

The second type of water production directly competes with oil production. This water usually flows to the production well by a path separate from that for oil (e.g. water coning, a high-permeability water channel through the oil strata, fractures, or wormholes, as in unconsolidated media). This type of water production can be more catastrophic to sustaining production and should be the target of conformance or water shut-off treatments. Conformance treatments such as inorganic, multi-valent crosslinked gels, microgels, and nanogels are typically applied at the injection well in an attempt to fix the flow anomaly from the source. Other methods such as relative permeability modifiers can be applied at the production well to prevent water production from entering the wellbore.

The main difficulty lies in identifying the origin of the excessive water production [76]. However, in the vast majority of cases, a single solution consisting of bullheading a crosslinked polymer gel is applied. As the name implies, associations form between the positive crosslinker species and negatively charged functional groups on the polymer backbone, effectively transitioning from a mobile fluid known as a *gelant* to a more structured species, the *gel*. There is an important difference between a gel injection treatment and polymer solution injection for EOR. For the former, the objective is to *minimize* penetration of gelants or permeability-reducing agents into the less-permeable, oil-productive zone. Any gel or blocking agent that enters the less-permeable zones can hinder (or even shut off) subsequent injected fluids (e.g. water) from entering and displacing oil from those zones (Figure 2.12). In contrast, polymer floods and similar mobility-control, EOR methods are intended to directly displace oil from less-permeable zones, and therefore their contact in these zones should be *maximized*. Several papers detail the applicability of both methods to help maximize the recovery from a given field [77–79].

2.2.5.2. Colloidal Dispersion Gels

Acrylamide-polymer gels formulated with low concentrations (200–1200 ppm) of polymer and aluminum citrate are referred to as CDGs. They are often used as large-volume treatments applied through injection wells to "matrix rock" for conformance improvement of waterflood sweep efficiency. Mechanisms and the means by which this particular gel technology generates incremental oil production are still controversial.

The vast majority of published laboratory studies have reported the following regarding aluminum-citrate CDGs [3, 80–84]:

- CDGs of acrylamide polymers crosslinked with aluminum citrate are not readily injectable into, and have difficulty

Comparison of gel treatment and polymer flooding. The gel treatment should only access the zones to plug while staying away from the unswept zones

Figure 2.12

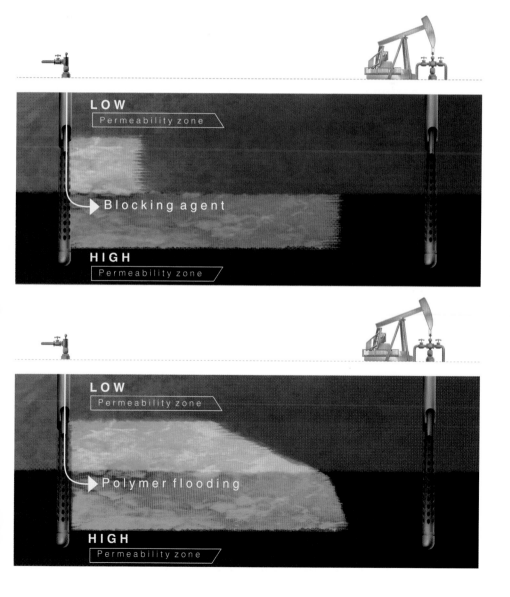

propagating through, matrix rock of normal permeabilities (e.g. sandstones of <1000 md). Seright discusses an experimental flooding study where an aluminum-citrate CDG was observed to not readily propagate, after two hours of aging, through matrix rock during the injection of an aluminum-citrate CDG solution, containing 300 ppm high-molecular weight partially hydrolyzed polyacrylamide (HPAM), into a 700-md Berea sandstone core plug at 105°F and at a superficial velocity through the sandstone of 16 ft./d [3].

- Aluminum does not readily propagate through sandstones and clay-containing rocks.

- Aluminum crosslinking of the polymer of CDGs normally occurs within several hours. This depends upon salinity, temperature, and pH.

- Aluminum-citrate CDGs do not preferentially enter high-permeability zones: this statement defies basic reservoir engineering laws, including Darcy's.

- Aluminum-citrate CDGs do not viscosify water any more than polymer alone, without the addition of the crosslinking chemical.

Given the poor laboratory results obtained with CDG injection, and the common problems associated with the technology, there are two possible explanations for the apparent success of a number of large-volume aluminum-citrate CDG treatments in terms of generating conformance improvement and incremental oil production when treating matrix-rock conformance problems [85–88]:

- The reservoir had some fractures or microfractures that allowed propagation and eventually were plugged by gel components.

- The high-permeability layers were in the multi-Darcy range, more accepting of the gel components for in-depth propagation.

Careful laboratory studies are still needed to validate the concept before going to a costly field deployment with significant operational expenditures. There is no credible laboratory, theoretical, or field evidence that CDGs can propagate deep into the porous rock of a reservoir and, at the same time, provide resistance factors (apparent viscosity in porous media) or residual resistance factors (permeability reduction) that are greater than those for the same polymer formulation without the crosslinker.

2.2.5.3. Microgels and Nanogels

Microgels and nanogels consist of particles that can adsorb on the pore wall and/or plug pore throats, therefore reducing the ability of water to flow through watered-out zones (decrease permeability). The products are often delivered in an inverse emulsion form and inverted in the field before injection. Several hurdles limit the deployment of such technologies:

- A good reservoir understanding on the micropore scale is required to select the best particle size and ensure a proper placement.

- Good product design, inversion, and deployment are required to minimize injectivity issues.

Two types of products are usually used. The first technology corresponds to calibrated particles with sizes ranging from 20 to 200 nm, sometimes up to a micron size, that will not expand much after release. The second technology corresponds to particles that can swell after a required time (delayed and swellable particles). The swelling ratio can vary from 5 to 20 times (sometimes up to 300 times the original size for super-absorbents), and the timing can be adjusted from several days to weeks depending on reservoir conditions (Figure 2.13).

The use of such technology requires a thorough reservoir understanding and characterization, especially the distribution of porosity, pore throat size, and connectivity. A careful laboratory design is also required to adapt the polymer based on reservoir temperature and salinity. It is important to consider injectivity tests in long and heterogeneous cores at both ambient and reservoir temperatures, to check differential retention and swelling kinetics for the swellable polymer particles.

Figure 2.13 # Delayed microgels – characteristics and principle of application

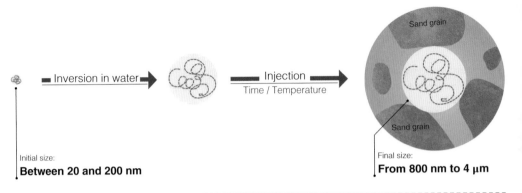

Inversion in water →

Injection →
Time / Temperature

Sand grain

Sand grain

Initial size:
Between 20 and 200 nm

Final size:
From 800 nm to 4 µm

When the product comes in an inverse emulsion form, a proper package is required to fully invert and release the polymer in the water phase. A bad inversion can lead to injectivity issues and poor particle propagation. When using a microgel or nanogel, a critical question is, how much damage or loss of flow capacity will these materials cause in oil-productive pathways in a reservoir?

2.2.5.4. Relative Permeability Modifiers (RPM)

The most commonly used relative permeability modifiers (RPMs) are solutions of water-soluble, high-molecular-weight polymers that adsorb on the pore walls and decrease the ability of water to flow. This type of treatment is normally reserved for production wells with no value for application to the injection side. The main issues are linked to candidate selection and the real impact on water and oil production. For the latter, decreasing the ability of water to flow will lead to an increase and buildup of water saturation at some radius away from the wellbore, increasing the relative permeability to water at the rear of the treatment and creating a *waterblock* zone. In the long term, it will therefore decrease the relative permeability to oil and impact the flow of the latter. An extensive review with discussion is given in Sydansk and Seright [89].

Chemical EOR techniques aim to improve sweep efficiency and recover trapped oil. ASP formulations require more careful design than straight polymer injection, are more costly to implement, and have different constraints, typically on the side of water quality, treatment, and rate of retention/adsorption of each chemical constituent. A subtle balance should be found between laboratory studies and field implementation to minimize cost and obtain relevant and reliable data that can be used to make the best engineering judgment when attempting to extrapolate to a larger scale.

Series, 6. Richardson, TX: Henry L. Doherty Memorial Fund of AIME, Society of Petroleum Engineers.

[2] Nutting, P.G. (1925). Chemical problems in the water driving petroleum from oil sands. *Industrial and Engineering Chemistry* 17: 1035–1036.

[3] New Mexico Petroleum Recovery Research Center. Reservoir sweep improvement. http://www.prrc.nmt.edu/groups/res-sweep.

[4] Fulin Y., Demin W., Xizhi Y. et al. 2004. High concentration polymer flooding is successful. Paper SPE 88454 presented at the SPE Asia Pacific Oil and Gas Conference and Exhibition, Perth, Australia, 18-20 October. https://doi.org/10.2118/88454-MS.

[5] Fulin Y., Demin W., Wenxiang W. et al. 2006. A pilot test of high-concentration polymer flooding to further enhance oil recovery. Paper SPE99354 presented at the SPE/DOE Symposium on Improved Oil Recovery, Tulsa, OK, USA, 22–26 April. https://doi.org/10.2118/99354-MS.

[6] Gogarty, W.B., Meabon, H.P., and Milton, H.W. (1970). Mobility control design of miscible-type waterfloods using micellar solutions. *Journal of Petroleum Technology* 141–147.

[7] Johnson, C.E. Jr. (1976). Status of caustic and emulsion methods. *Journal of Petroleum Technology* 85–92.

[8] Hirasaki, G.J. (1981). Application of the theory of multicomponent, multiphase displacement to three-component, two-phase surfactant flooding. *Society of Petroleum Engineers Journal* 191–204.

[9] Han, D. (2001). *Surfactant Flooding: Principles and Applications*. Beijing, China: Petroleum Industry Press.

[10] Schramm, L. (2000). *Surfactants: Fundamentals and Applications in the Petroleum Industry*, 621. Cambridge University Press.

[11] Hirasaki, G.J., Miller, C.A., and Puerto, M. (2011). Recent advances in surfactant EOR. *Society of Petroleum Engineers Journal* 16: 889–907.

[12] Levitt, D., A.C. Jackson, C. Heinson et al. 2006. Identification and evaluation of high-performance EOR surfactants. SPE/DOE Symposium on Improved Oil Recovery, Tulsa, Oklahoma, USA.

[13] Zhao, P., A.C. Jackson, C. Britton et al. 2008. Development of high-performance surfactants for difficult oils. Paper SPE 113432 presented at the SPE Symposium on Improved Oil Recovery, Tulsa, Oklahoma, USA, 20-23 April. https://doi.org/10.2118/113432-MS.

[14] Barnes, J.R., Smit, J.P., Smit, J.R. et al. 2008. Development of surfactants for chemical flooding at difficult reservoir conditions. Paper SPE113313 presented at the SPE Symposium on Improved Oil Recovery, Tulsa, Oklahoma, USA, 20-23 April https://doi.org/10.2118/113313-MS.

[15] Stegemeier, G.L. (1977). Mechanisms of entrapment and mobilization of oil in porous media. In: *Improved Oil Recovery by Surfactant and Polymer Flooding* (ed. D.O. Shah and R.S. Schechter), 55–91. New York: Academic Press.

[16] Koval, E.J. (1963). A method for predicting the performance of unstable miscible displacement in heterogeneous media. *Society of Petroleum Engineers Journal* 3 (2): 145–154.

[17] Kahlweit, M., Strey, R., Firman, P. et al. (1988). General patterns of the phase behavior of mixtures of water, nonpolar solvents, amphiphiles, and electrolytes. *Langmuir* 4 (3): 499–511. https://doi.org/10.1021/la00081a002.

[18] Southwick, G.J., Svec, Y., Chilek, G. et al. 2010. The effect of live crude on alkaline-surfactant-polymer formulations: implications for final formulation design. Paper SPE 135357 presented at the SPE Annual Technical Conference and Exhibition, Florence, Italy, 19-22 September. https://doi.org/10.2118/135357-PA.

[19] Puskas S., Vago A., Toro M. et al. 2017. First surfactant-polymer EOR injectivity test in the Algyo field, Hungary. Paper EAGE ThB08 presented at the 19th European Symposium on Improved Oil Recovery, Stavanger, Norway, 24-27 April.

[20] Zhu, Y.-Y., Hou, Q.F., Liu, W.D. et al. 2012. Recent progress and effects analysis of ASP flooding field tests. Paper SPE 151285 presented at the SPE Improved Oil Recovery Symposium, Tulsa, Oklahoma, USA, 14-18 April. https://doi.org/10.2118/0113-0077-JPT.

[21] Al-Amrie, O., Peltier, S., Pearce, A. et al. 2015. The first successful chemical EOR pilot in the UAE: one spot pilot in high temperature, high salinity carbonate reservoir. Paper SPE177514 presented at the Abu Dhabi International Petroleum Exhibition and Conference, Abu Dhabi, UAE, 9-12 November. https://doi.org/10.2118/177514-MS.

[22] Nelson, R.C. (1983). The effect of live crude on phase behavior and oil–recovery efficiency of surfactant flooding systems. *Society of Petroleum Engineers Journal* 501–510.

[23] Nelson, R.C., Lawson, J.B., Thigpen, D.R. et al. 1984. Cosurfactant-enhanced alkaline flooding. Paper SPE 12672 presented at the SPE/DOE Fourth Symposium on Enhanced Oil Recovery, Tulsa, Oklahoma, 15.18 April.

[24] Dean, M. 2011. Selection and evaluation of surfactants for field pilots. Masters thesis. University of Texas at Austin.

[25] Sheng, J.J. (2011). *Modern Chemical Enhanced Oil Recovery: Theory and Practice.* Elsevier.

[26] Sharma, A, Azizi-Yarand, A., Clayton, B. et al. 2012. The design and execution of an alkaline-surfactant-polymer pilot test. Paper SPE 154318 presented at the SPE Improved Oil Recovery Symposium, Tulsa, Oklahoma. 14-18 April. https://doi.org/10.2118/154318-PA.

[27] Cheng, K.H. 1986. Chemical consumption during alkaline flooding: a comparative evaluation. Paper SPE 14944 presented at the SPE/DOE Symposium on Enhanced Oil Recovery, Tulsa, Oklahoma, 20-23 April. https://doi.org/10.2118/14944-MS.

[28] Sheng J., 2013. A Comprehensive Review of ASP Flooding. Paper SPE165358 presented at the SPE Western Regional & AAPG Pacific Section Meeting, 2013 Joint Technical Conference held in Monterey, California, USA, 19−25 April. https://doi.org/10.2118/165358-MS.

[29] Zubari, H.K. and Sivakumar, V.C.B. 2003. Single well tests to determine the efficiency of alkaline-surfactant injection in a highly oil-wet limestone reservoir. Paper SPE81464 presented at the Middle-East Oil Show, Bahrain, 9–12 June.

[30] Stoll, W.M., al Shureqi, H., Final, J. et al. (2011). Alkaline/surfactant/polymer flood: from the laboratory to the field. *SPE Reservoir Evaluation & Engineering* 14 (5): 702–712.

[31] Weatherill, A. 2009. Surface development aspects of alkali-surfactant-polymer (ASP) flooding. Paper IPTC 13397 presented at the International Petroleum Technology Conference, Doha, Qatar, 7-9 December. https://doi.org/10.2523/IPTC-13397-MS.

[32] Wang, C.-L., Wang, B.-Y., Cao, X.-L. et al. 1997. Application and design of alkaline–surfactant–polymer system to close well spacing pilot Gudong oilfield. Paper SPE 38321 presented at the SPE Western Regional Meeting, Long Beach, California, June 25-27. https://doi.org/10.2118/38321-MS.

[33] Gao, S.-T., Li, L.-H., Yang, Z.-Y. et al. (1996). Alkaline/surfactant/polymer pilot performance of the west central Saertu, Daqing oil field. *SPE Reservoir Evaluation & Engineering* 11 (3): 181–188.

[34] Li, H.-B., Li, H.-F., Yang, Z.-Y. et al. (1999). ASP pilot test in Daqing Sa-Zhong-Xi region. *Oil and Gas Recovery Technology* 6 (2): 15–19.

[35] Wang, H.-Z., Liao, G.-Z., and Song, J. (2006). Combined chemical flooding technologies. In: *Technological Developments in Enhanced Oil Recovery* (ed. P.P. Shen), 126–188. Beijing, China: Petroleum Industry Press.

[36] Wang, D.-M., Cheng, J.-C., Wu, J.-Z. et al. 1998. An alkaline/surfactant/polymer field test in a reservoir with a long-term 100% water cut. Paper SPE 49018 presented at the SPE Annual Technical Conference and Exhibition, New Orleans, Louisiana, 27-30 September. https://doi.org/10.2118/49018-MS.

[37] Wang, D.-M., Cheng, J.-C., Li, Q. et al. 1999a. An alkaline bio-surfactant polymer flooding pilots in Daqing oil field. Paper SPE 57304 presented at the SPE Asia Pacific Improved Oil Recovery Conference, Kuala Lumpur, Malaysia, 25-26 May. https://doi.org/10.2118/57304-MS.

[38] Li, H.-F., Liao, G. Z., Han P.-H. et al. 2003. Alkaline/surfactant/polymer (ASP) commercial flooding test in the central Xing2 area of Daqing oilfield. Paper SPE 84896 presented at the International Improved Oil Recovery Conference, Kuala Lumpur, Malaysia, 20-21 October. https://doi.org/10.2118/84896-MS.

[39] Qu, Z.-J., Zhang, Y.-G., Zhang, X.-S. et al. 1998. A successful ASP flooding pilot in Gudong oil field. Paper SPE/DOE 39613 presented at Improved Oil Recovery Symposium, Tulsa, Oklahoma, 19-22 April. https://doi.org/10.2118/39613-MS.

[40] Song, W.-C., Yang, C.-Z., Han, D.-K. et al. 1995. Alkaline–surfactant–polymer combination flooding for improving recovery of the oil with high acid value. Paper SPE 29905 presented at the International Meeting on Petroleum Engineering, Beijing, China, 14-17 November. https://doi.org/10.2118/29905-MS.

[41] Yang, Z.-J., Zhu, Y., Ma, X.-Y. et al. (2002). ASP pilot test in Guodao field. *Journal of Jianghan petroleum Institute* 24 (1): 62–64.

[42] Cao, X.-L., Sun, H.-Q., Jiang, Y.-B. et al. (2002). ASP pilot test in the Western Guodao field. *Oilfield Chemistry* 19 (4): 350–353.

[43] Chang, H.L., Zhang, Z.-Q., Wang, Q.-M. et al. (2006). Advances in polymer flooding and alkaline/surfactant/polymer processes as developed and applied in the People's Republic of China. *Journal of Petroleum Technology* 84–89.

[44] Gu, H.-J., Yang, R.-Q., Guo, S.G. et al. 1998. Study on reservoir engineering: ASP flooding pilot test in Karamay oilfield. Paper SPE 50918 presented at the International Oil and Gas Conference and Exhibition, Beijing, China, 2-6 November. https://doi.org/10.2118/50918-MS.

[45] Delshad, M., Han, W., Pope, G.A. et al. 1998. Alkaline/surfactant/polymer flood predictions for the Karamay oil field. Paper SPE 39610 presented at the SPE/DOE Improved Oil Recovery Symposium, Tulsa, Oklahoma, 19-22 April. https://doi.org/10.2118/0199-0034-JPT.

[46] Qiao, Q., Gu H.-J., Li D.-W. et al. 2000. The pilot test of ASP combination flooding in Karamay oil field. Paper SPE 64726 presented at the International Oil and Gas Conference and Exhibition, Beijing, China, 7-10 November. https://doi.org/10.2118/64726-MS.

[47] Zhang, Z.-L., Yang, Y.-M., Hong, L. et al. (2001). ASP pilot test in Honggang field. *Journal of Southwest Petroleum Institute* 23 (3): 47–9, 57.

[48] Jiang, J.-L., Guo, D.-F., Li, X.-F. et al. (2003). Pilot field trial of natural mixed carboxylates/xanthan flood at well H5–15 block in Huzhuangli oil field. *Oilfield Chemistry* 20 (1): 58–60.

[49] Wang, D.-C., Yang, T.-R., Du, T.-P. et al. (1999). Micellar/polymer flooding pilot test in the H184 well pattern in the Laojunmiao field. *Petroleum Exploration and Development* 26 (1): 47–49.

[50] Vargo, J., Turner, J., Vergnani, B. et al. (2000). Alkaline-surfactant-polymer flooding of the Cambridge Minnelusa field. *SPE Reservoir Evaluation & Engineering* 3 (6): 552–558.

[51] Clark, S.R., Pitts, M.J., and Smith, S.M. (1993). Design and application of an alkaline-surfactant-polymer recovery system to the west Kiehl field. *SPE Advanced Technology Series* 1 (1): 172–179.

[52] Meyers, J.J., Pitts, M.J., and Wyatt, K. 1992. Alkaline-surfactant-polymer flood of the West Kiehl, Minnelusa unit. Paper SPE 24244 presented at the SPE/DOE Enhanced Oil Recovery Symposium, Tulsa, Oklahoma 22–24 April. https://doi.org/10.2118/24144-MS.

[53] Pitts, M.J., Dowling, P., Wyatt, K. et al. 2006. Alkaline-surfactant-polymer flood of the Tanner field. Paper SPE 100004 presented at the SPE/DOE Symposium on Improved Oil Recovery, Tulsa, Oklahoma, 22–26 April. https://doi.org/10.2118/100004-MS.

[54] www.surtek.com

[55] McInnis L., Hunter K., Ellis-Toddington T. et al. 2013. Case study of the Mannville B ASP flood. Paper SPE165264 presented at the SPE Enhanced Oil Recovery Conference, Kuala Lumpur, Malaysia, 2–4 July. https://doi.org/10.2118/165264-MS.

[56] Zargon Oil & Gas Ltd. (2017). Corporate presentation.http://zargon.ca/wp-content/uploads/2017/08/Zargon-Corporate-Presentation-august-9-2017-rev-6-1.pdf.

[57] Watson A., Trahan G.A., and Sorensen W. 2014. An interim case study of an alkaline-surfactant-polymer flood in the Mooney field, Alberta, Canada. Paper SPE 169154 presented at the SPE Improved Oil Recovery Symposium, Tulsa, Oklahoma, USA, 12–16 April. https://doi.org/10.2118/169154-MS.

[58] Liu J.Z., Adegbesan K., and Bai J.J.. 2012. Suffield area, Alberta, Canada – Caen polymer flood pilot project. Paper SPE 157796 presented at the SPE Heavy Oil Conference, Calgary, Alberta, Canada, 12–14 June. https://doi.org/10.2118/157796-MS.

[59] Hernandez, C., Chacon, L.J., Anselmi, L. et al. 2002. Single well chemical tracer test to determine ASP injection efficiency at Lagomar VLA-6/9/21 area, C4 member, Lake Maracaibo, Venezuela. Paper SPE 75122 presented at the SPE/DOE Improved Oil Recovery Symposium, Tulsa, Oklahoma, USA, 13–17 April. https://doi.org/10.2118/75122-MS.

[60] Hernandez, C., Chacon, L.J., Anselmi, L. et al. (2003). ASP system design for an offshore application in La Salina field, Lake Maracaibo. *SPE Reservoir Evaluation & Engineering* 6 (3): 147–156.

[61] Manrique, E., De Carvajal, G., Anselmi, L. et al. 2000. Alkali/surfactant/polymer at VLA 6/9/21 field in Maracaibo Lake: experimental results and pilot project design. Paper SPE 59363 presented at the SPE/DOE Improved Oil Recovery Symposium, 3–5 April, Tulsa, Oklahoma, USA. https://doi.org/10.2118/59363-MS.

[62] Pratap, M. and Gauma, M.S. 2004. Field implementation of alkaline-surfactant-polymer (ASP) flooding, a maiden effort in India. Paper SPE 88455 presented at the SPE Asia Pacific Oil and Gas Conference and Exhibition, Perth, Australia, 18–20 October. https://doi.org/10.2118/88455-MS.

[63] Jain, A.K., Dhawan, A.K., and Misra, T.R. 2012. ASP flood pilot in Jhalora (K-IV), India - a case study. Paper SPE 153667 presented at the SPE Oil and Gas India Conference and Exhibition, Mumbai, India, 28–30 March. https://doi.org/10.2118/153667-MS.

[64] Pandey A., Koduru N., Stanley M. et al. 2016. Results of ASP pilot in Mangala field: a success story. Paper SPE179700 presented at the SPE Improved Oil Recovery Conference, Tulsa, Oklahoma, USA, 11–13 April. https://doi.org/10.2118/179700-MS.

[65] Al-Shuaili, K., Guntupalli, S., Al-Amri, M. et al. 2016. Alkali-surfactant-polymer pilot implementation in South Oman: establishing waterflood baseline. Paper SPE 179752 presented at the SPE EOR Conference at Oil and Gas, Muscat, Oman, 21–23 March. https://doi.org/10.2118/179752-MS.

[66] Van der Heyden, F.H.J., Mikhaylenko, E., de Reus, A.J. et al. 2017. Injectivity experiences and its surveillance in the west Salym ASP pilot. Paper EAGE ThB07 presented at the 19th European Symposium on Improved Oil Recovery, Stavanger, Norway, 24-27 April.

[67] Chai, C.F., Adamson, G., Lo, S.W. et al.2011. Paper SPE 144594 presented at the SPE Enhanced Oil Recovery Conference, Kuala Lumpur, Malaysia, 19–21 July. https://doi.org/10.2118/144594-MS.

[68] Moreno, R., Anselmi, L., Coombe, D. et al. 2003. Comparative mechanistic simulations to design an ASP field pilot in La Salina, Venezuela. Paper CIM-199 presented at the Canadian International Petroleum Conference, Calgary, Alberta, Canada, 10–12 June.

[69] Guerra, E., Valero, E., Rodríguez, D. et al. 2007. Improved ASP design using organic compound-surfactant-polymer (OCSP) for La Salina field, Maracaibo Lake. Paper SPE 107776 presented at the Latin American & Caribbean Petroleum Engineering Conference, Buenos Aires, Argentina, 15–18 April. https://doi.org/10.2118/107776-MS.

[70] Olsen, D.K., Hicks, M.D., Hurd, B.G. et al. 1990. Design of a novel flooding system for an oil-wet central Texas carbonate reservoir. Paper SPE 20224 presented at the SPE/DOE Seventh Symposium on Enhanced Oil Recovery, Tulsa, Oklahoma, USA, 22–25 April. https://doi.org/10.2118/20224-MS.

[71] Yang, H.D. and Wadleigh, E.E. 2000. Dilute surfactant IOR – design improvement for massive, fractured carbonate applications. Paper SPE 59009 presented at the SPE International Petroleum Conference and Exhibition, Villahermosa, Mexico, 1-3 February. https://doi.org/10.2118/59009-MS.

[72] Rilian, N.A., Sumestry, M., and Wahyuningsih. 2010. Surfactant stimulation to increase reserves in carbonate reservoir -- A case study in Semoga field. Paper SPE 130060 presented at the SPE EUROPEC/EAGE Annual Conference and Exhibition, Barcelona, Spain, 14–17 June, https://doi.org/10.2118/130060-MS.

[73] Sheng, J.J. (2013). Surfactant enhanced oil recovery in carbonate reservoirs. *Advances in Petroleum Exploration and Development* 6 (1): 1–10.

[74] Xie, X., Weiss, W.W., Tong, Z., and Morrow, N.R. (2005). Improved oil recovery from carbonate reservoirs by chemical stimulation. *Society of Petroleum Engineers Journal* 276–285.

[75] Weiss, W.W., Xie, X., Weiss, J. et al. (2006). Artificial intelligence used to evaluate 23 single-well surfactant-soak treatments. *SPE Reservoir Evaluation & Engineering* 209–216.

[76] Seright R.S., Lane RH., and Sydansk R.D., 2003. A strategy for attacking excess water production. Paper SPE84966, SPE Production & Facilities, August. https://doi.org/10.2118/84966-PA.

[77] Seright R.S., 1996. Improved techniques for fluid diversion in oil recovery processes. Final report, DOE/BC/14880–15, 62–89. US DOE.

[78] Seright, R.S. 1997. Improved methods for water shut-off. Semiannual report, DOE/PC/91008–1, 82–95.

[79] Seright, R.S. 1995. Gel placement in fractured systems. SPEPF 241; Trans. AIME, 299.

[80] Smith J.E., Liu H., and Guo Z.D. 2000. Laboratory studies of in-depth colloidal dispersion gel technology for daqing oil field. . Paper SPE 62610 presented at the 2000 SPE/AAPG Western Regional Meeting, Long Beach, CA, USA, 19–23 June. https://doi.org/10.2118/89460-MS.

[81] Rocha C.A., Green D.W., Willhite G.P. et al. 1989. An experimental study of the interactions of aluminum citrate solutions and silica sand. Paper SPE 18503 presented at the SPE International Symposium on Oilfield Chemistry, Houston, TX, USA, 8–10 February. https://doi.org/10.2118/18503-MS.

[82] Ranganathan, R., Lewis, R., McCool, C.S. et al. (1998). Experimental study of the gelation behavior of a polyacrylamide/Aluminum citrate colloidal-dispersion gel system. *Society of Petroleum Engineers Journal* 337–343.

[83] Walsh M.P., Rouse B.A., Senol N.N. et al. 1983. Chemical interactions of aluminum-citrate solutions with formation minerals. Paper SPE 11799 presented at the SPE International Symposium on Oilfield Chemistry, Denver, CO, USA, 1–3 June. https://doi.org/10.2118/11799-MS.

[84] Stavland A. and Johsbraten H.C. 1996. New insight into aluminum citrate/poly-acrylamide gels for fluid control. Paper SPE/DOE 35381 presented at the SPE/DOE Symposium on Improved Oil Recovery, Tulsa, Oklahoma, USA, 21–24 April. https://doi.org/10.2118/35381-MS.

[85] Chang H.L., Sui X., Xiao L. et al. 2004. Successful field pilot of in-depth colloidal dispersion gel (cdg) technology in Daqing oil field. Paper SPE89460 presented at the SPE/DOE Symposium on Improved Oil Recovery, Tulsa, Oklahoma, USA, 17–21 April. https://doi.org/10.2118/89460-MS.

[86] Diaz D., Somaruga C., Norman C. et al. 2008. Colloidal dispersion gels improve oil recovery in a heterogeneous argentina waterflood. Paper SPE113320 presented at the SPE/DOE Improved Oil Recovery Symposium, Tulsa, Oklahoma, USA, 19–23 April. https://doi.org/10.2118/113320-MS.

[87] Manrique E., Reyes S., Romero J. et al. 2014. Colloidal dispersion gels (CDG): field projects review. Paper SPE169705 presented at the SPE EOR Conference at Oil and Gas West Asia, Muscat, Oman, 31 March – 2 April. https://doi.org/10.2118/169705-MS.

[88] Castro R., Maya G., Sandoval J., Leon J. et al. 2013. Colloidal dispersion gels (CDG) in Dina cretaceos field: from pilot design to field implementation and performance. Paper SPE165273 presented at the SPE Enhanced Oil Recovery Conference, Kuala Lumpur, Malaysia, 2–4 July. https://doi.org/10.2118/165273-MS.

[89] Sydansk R.D. and Seright R.S. 2006. When and where relative permeability modification water shutoff treatments can be successfully applied. Paper SPE99371 presented at the SPE/DOE Symposium on Improved Oil Recovery, Tulsa, Oklahoma, USA, 22–26 April. https://doi.org/10.2118/99371-MS.

Chapter three

Polymer Flooding

Viscous water injection, also known as polymer flooding (PF), will be the focus of this chapter. The technological concept will be discussed along with general concerns raised by the injection of viscous water.

Essentials of Polymer Flooding Technique, First Edition. Antoine Thomas.
© 2019 John Wiley & Sons Ltd. Published 2019 by John Wiley & Sons Ltd.

3.1. Introduction

Among chemical enhanced oil recovery methods (CEOR), polymer flooding (PF) is an intuitive technique with a long commercial history and many field-proven results. During recovery of oil with a variety of viscosities, it's only natural to consider a recovery process in which injection water is viscosified. The envelope of PF application has widened greatly over the years thanks to numerous improvements in chemistry, formulation, design, and implementation. Polymer flooding consists of injecting polymer-augmented water that exhibits an increased fluid viscosity, improving the sweep efficiency of the reservoir flood by providing mobility control between water and resident hydrocarbons.

The first significant commercial uses of polymers to improve the oil recovery factor of oil reservoirs can be traced back to the 1970s in the United States during a crude oil price-control period. At that time, economic incentive programs played an important role in the rapid development and implementation of CEOR processes. Direct financial support was provided by the Energy Research and Development Administration (ERDA) and the Department of Energy (DOE) from 1974 to 1980 through cost-sharing programs. Between 1979 and 1981, the Department of Energy-Enhanced Oil Recovery (DOE-EOR) incentive program allowed oil producers to sell oil at higher prices, while the Crude Oil Windfall Profit Tax Act released in 1980 reduced tax rates for enhanced oil recovery (EOR) projects. The Economic Recovery Tax Act of 1981 supported research and experimentation through tax credits. High oil prices created a push to implement polymer EOR projects that may not have been correctly designed from a scientific standpoint, and therefore may have appeared economically successful but were possibly technical failures in terms of how much oil remained unrecovered. For polymer flooding, this translated into very small polymer concentrations and bank sizes that were totally inefficient at improving reservoir sweep efficiency and oil recovery factors [1].

In the 1980s, the number of projects dramatically decreased, mainly due to low oil prices and poor understanding of this particular EOR technique. Research continued over the years, and interest in polymer injection was renewed after the Daqing oil field in China starting polymer injection in the 1990s. Many projects were successfully launched after 2000 in Canada, the Middle East, India, and South America. The first offshore trials were conducted in 2010; and, today, polymer injection is implemented after primary production in many fields without prior water injection.

3.2. Concept

Evidence that a difference in fluid mobilities would affect waterflood performance was first recognized by Muskat in 1949 [2]. Stiles [3], in 1950, used permeability and capacity distribution in waterflood calculations, while Dykstra and Parsons [4] showed the impact of mobility ratio and vertical permeability variations on oil recovery. The concept of mobility ratio was discussed later by Aronofsky and Ramey [5] in 1956 along with its influence on flood patterns and injection/production history in a five-spot pattern. In 1954, Dyes et al. [6] presented studies of the effect of mobility ratio on oil production after water breakthrough. These studies led to the conclusion that increasing water viscosity could greatly improve sweep efficiency. Research work by Pye [7] and Sandiford [8] showed that adding water-soluble polymers to the injected brine could effectively reduce the mobility of water. Many studies have since been performed dealing with in situ polymer behavior and rheology.

3.2.1. Fractional Flow

It is very important for a petroleum engineer to properly estimate flooding parameters such as oil rate, the volume of oil displaced at any time, and the water volumes that will be handled. Several simple tools are available to perform calculations and build business cases [9–13]. Fractional flow theory has always assisted engineers in designing waterfloods but can also greatly help understand EOR processes, especially polymer floods. This method assumes that (i) a flood front exists, (ii) oil and water move behind the front, and (iii) water doesn't move ahead of the front. Given the Buckley-Leverett theory for a steady-state, one-dimensional, two-immiscible, incompressible phase displacement, neglecting gravity and capillary pressure, the fractional flow (f_w) can be defined as:

$$f_w = \frac{q_w}{q_o + q_w}$$

where q_w and q_o are the volumetric flow rates of water and oil at reservoir conditions (defined using Darcy's law).

Each q in the equation can be replaced using Darcy's law; and, assuming the system's length, area, and pressure drops are constant, the previous equation reduces to:

$$f_w = \frac{1}{1 + \dfrac{k_{ro}(S_{wc}) * \mu_w}{k_{rw}(S_{or}) * \mu_o}}$$

Recall from Chapter 2 the end-point mobility ratio

$$\mathbf{M} = \frac{\lambda_o}{\lambda_w} = \frac{\mu_o \, / \, \mathbf{kro(Swc)}}{\mu_w \, / \, \mathbf{krw(Sor)}}$$

where λ, μ, and k are the mobility, viscosity, and effective end-point permeabilities, respectively, and where the subscripts w and o refer to water and oil.

The final equation form becomes:

$$f_w = \frac{1}{1 + \dfrac{1}{M}}$$

The fractional flow of water can be easily plotted as a function of water saturation. It is possible to determine graphically the average water saturation $S_{w\,avg}$ behind the flood front at break-through by drawing a tangent to the curve $f_w = f(S_w)$ through the initial water saturation S_{wi} (Figure 3.1). From these calculations, it is also possible to determine cumulative injected water, oil recovery, and breakthrough time [9–11].

Figure 3.1

Fractional flow curves for three mobility ratios and corresponding tangents to the curves

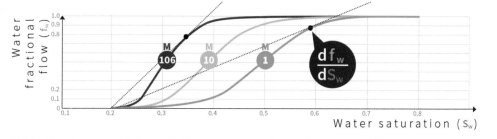

Examples of fractional flow curves with different mobility ratios are given in Figure 3.1. The water saturation at the front is also given for a high mobility ratio case and a low mobility ratio case (red arrows).

Given the following equation to calculate the breakthrough time, it is possible to see just by looking at the slope of the tangent that the highest mobility ratio gives a much earlier breakthrough time. The same approach can be considered to calculate the volume of oil recovered at breakthrough: it is much higher for the low mobility ratio case.

$$t_{BT} = PV * i_w * \left(\frac{1}{\frac{df_w}{dS_w}} \right)_{Swf}$$

where t_{BT} is the breakthrough time, PV is the reservoir pore volume, i_w is the water injection rate, and the value of the slope at breakthrough is $\left(\dfrac{df_w}{dS_w}\right)_{Swf}$.

The idea behind this simple, rough mathematical consideration is to show the value of decreasing the mobility ratio by increasing water viscosity via polymer addition.

3.2.2. Polymer Flooding Applicability

Oil is left behind in a waterflood either because it is trapped by the capillary forces (in the form of discontinuous residual oil) or because it is bypassed [13]. Improving the mobility ratio with the use of polymers minimizes this bypassing effect. Polymer flooding is normally implemented in two cases [14]:

- When the mobility ratio during a waterflood is not favorable, increasing water viscosity through polymer addition can increase sweep efficiency and oil recovery (see Section 3.2.1).

- Even with a favorable mobility ratio, if the reservoir is heterogeneous, polymer injection can help reduce water mobility in the high-permeability layers to push the oil from the low-permeability layers.

In the first case, inefficient microscopic displacement leads to early water breakthrough followed by a long period of two-phase production with increasing water-cut. This situation can be illustrated by the *viscous fingering* concept, which occurs especially in heavy oil reservoirs or when the mobility ratio M is greatly above 1 (Figure 3.2).

Interestingly, the second case is very often overlooked. Even if the mobility ratio is favorable (around or even below 1), the presence of high-permeability channels, heterogeneities, or large-scale reservoir layering can greatly impair volumetric sweep efficiency during a secondary waterflood (it can also be

Flow profiles for two injection cases – unfavorable mobility ratio at the top, and favorable mobility ratio at the bottom

Figure 3.2

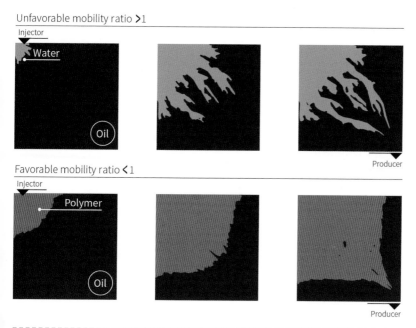

Unfavorable mobility ratio > 1

Favorable mobility ratio < 1

---

seen when applying fractional flow to a multilayered reservoir). A close look at the geological model and production history should help identify issues explaining the poor recovery factor obtained even when the mobility ratio is favorable. More precise screening criteria will be given later in this chapter.

3.2.3. Timing

As discussed in Chapters 1 and 2, EOR techniques should be considered at the very beginning of field development, to accelerate and maximize oil recovery. It is indeed paradoxical that, when it comes to hydrocarbon production in the oil and gas industry, finding a cure or solution to the problem is considered better than prevention. Knowing that, on average, 65% of the liquid hydrocarbons remain in reservoirs after secondary production, we can ask, "Why do engineers wait for the water-cut to reach uneconomical values before starting to look at techniques to improve oil recovery?" Possible explanations to this question are as follows:

- Lack of knowledge of the reservoir (geometry, fluid distribution, etc.). However, this lack of knowledge should be balanced with an approach grounded in science that may be somewhat intuitive: if the reservoir is highly heterogeneous and/or water is more mobile than oil due to a large viscosity contrast, there is a high probability that the waterflood will be inefficient.

- Definition of a baseline. Without upfront confidence in a technique, it is imperative to define a baseline to prove its efficiency. What is the baseline if EOR is started in a greenfield? Can it be called EOR? This question also has several answers. One is pragmatic: looking at the history of a given technology should help build trust and allow a proper design based on previous experience. A second answer would be to consider, when possible, several patterns to compare a regular scheme with primary, secondary, and tertiary recoveries and straight EOR techniques.

- Team changes. When considering the development of an oil field and the different production stages, including EOR, it is important to keep people with specific technical skills and to brief newcomers with a complete record of the project. Knowledge transfer within the team ensures that everyone understands the goals and potential technical challenges at each stage of implementation. Changing teams too often requires that individuals rebuild their technical skills; and knowing they may change position again can affect their motivation toward a new project. This can easily sidetrack or delay project progression.

For polymer injection, as for any EOR technique, the guideline is simple: the earlier the better. Some people in the industry have another definition of the acronym EOR: Early *Or* Regret. Long-term economics will always be better when EOR techniques are implemented early in the life of the field, thus accelerating and maximizing oil production while delaying water breakthrough and the costs associated with separation and treatment.

Injecting viscous polymer solutions at a high water-cut is not necessarily detrimental to the overall success of the project. However, it can complicate the implementation and interpretation of the results. After extensive water injection, the presence of water fingers can decrease the efficiency of polymer injection since the viscous fluid can still preferentially enter these zones, leading to early polymer breakthrough in some cases. The response time and changes in water-cut can also be delayed given the high water saturation in the layers swept preferentially during the secondary flood: water that will be pushed by the polymer slug.

3.3. Envelope of Application

3.3.1. History

The technical aspects of polymer injection continue to be studied extensively in the industry. Two aspects are usually considered: (i) polymer selection and stability and (ii) reservoir applicability. Literature review from the 1960s through the present day provides an interesting view of how the technology has evolved over the years.

In the 1960s and 1970s, the papers published clearly outlined a restricted application window for polymers and polymer flooding (Table 3.1). For example, in 1978, Chang [15] detailed some screening criteria for polymer flooding.

The maximum temperature considered at the time was 70°C for polyacrylamides and 95°C for xanthans. Interestingly, the

Table 3.1 — Envelope of application for polymer injection in 1978 [15]

Reservoir temperature (°C)	<95
Polyacrylamide	<70
Xanthan gums	<95
Crude oil viscosity (cP)	<200
Water-oil mobility ratio	>1
Mobile oil saturation (%PV)	>10
water-oil ratio (WOR)	<15 preferred
Average reservoir permeability (mD)	>20
Lithology	Sandstone preferred

Reservoirs with strong natural water drive, large gas caps, gross channeling, or major natural fractures should be avoided.

maximum value for crude oil viscosity was defined as 200 cP. Seright [16] pointed out two possible explanations for this particular value: (i) given the oil price at that time, 200 cP was viewed as the most viscous oil that could be economically recovered using polymers; and (ii) important injectivity decreases were expected during injection of viscous solutions, viscous enough to reach reasonable mobility ratios. A third possible explanation is that many viscous oil reservoirs were not water-flooded at this time; this approach was not rapidly deployed until the advent of horizontal well technology in the 1990s.

In Sorbie's 1991 book *Polymer-Improved Oil Recovery,* [14], other screening criteria were discussed as shown in Table 3.2. The same remarks apply to that case: the maximum oil viscosity was defined below 100 cP, with relatively low reservoir temperatures.

Screening criteria after Sorbie [14]

Table 3.2

Screening criterion	Viscosity control polymer flood	Heterogeneity control polymer flood	Comment
Oil viscosity	Usually $5\,cP < \mu_o < 30\,cP$ Max 70 cP	Usually $0.4\,cP < \mu_o < 10\,cP$ Max 20 cP	Symptom in both cases is early water break-through and low sweep efficiency
Level of large-scale heterogeneity	Low formation should be as homogeneous as possible	Some heteroge-neity by definition $4 < kh_i/k_{av} < 30$	For heterogeneity control, less severe contrast does not require polymer, and more severe is too high for normal polymer
Absolute permeability	>20 mD		To avoid excessive polymer retention
Temperature	Lower temperature best Best <80 °C Max <95 °C		Polymers degrade at higher temperatures.
Water injectivity	Should be good, preferably with some spare injection capacity – fracking may help		If there are problems with water, they will be worse with polymer.
Aquifer/oil/water contact	Injection not deep in aquifer or far below oil/water contact		Additional retention losses in transport to oil leg
Clays	Should be generally low		Tend to give high polymer retention
Injection brine salinity/hardness	Not critical but determines which polymer can be used		High salinity/hardness biopolymer Low salinity/hardness = poly-acrylamide (PAM)

Over the years, many improvements have helped widen the envelope of application for polymers in chemical EOR. First are those related to the polymers themselves, their chemistry, and their resistance to salinity and temperature. The incorporation of new monomers along with improvements in manufacturing processes, polymerization, and quality control of raw materials have opened the door to new reservoirs and harsher conditions. For instance, the use of monomers such as acrylamido tertiary butyl sulfonic acid (ATBS) and/or N-Vinylpyrrolidone (NVP) increases polymer resistance to salinity and temperature, taking the limit up to 120 °C in harsh brines (up to 120 g l^{-1}) [17, 18]. Some examples will be given later in the book.

Regarding reservoir characteristics, many field examples have pushed past the prior envelope of application: polymer injection has been implemented successfully in a reservoir with live oil viscosity up to 10 000 cP but also as low as 0.5 cP [19]. Field tests have been conducted at high temperature in carbonates (120 °C) and in low-permeability formations (3 mD). The current limits of polymers can be summarized as shown in Table 3.3 [1].

The salinity criterion should be analyzed carefully, taking into consideration the divalent content, which can be the most detrimental to polymer stability and may impact retention behavior.

Table 3.3

Current envelope of application for polymers in EOR

Parameter	Today
Oil viscosity	<10 000 cP
Temperature	<140 °C
Permeability	>10 mD
Salinity	<250 g l^{-1} TDS

3.3.2. Reservoir Prescreening

Given the criteria discussed in the previous section, it is possible to quickly screen reservoirs in a portfolio to discard those outside the limits given in Table 3.3. This step consists of selecting a zone or pattern to prove the concept and build an economic case by gathering all relevant data. The attempt of this section is to provide and explain a quick screening tool that selects the best reservoir candidate for polymer injection.

A list of parameters to rank mature reservoirs is given in Table 3.4.

Table 3.4

Parameters to consider when screening candidates for polymer flooding

Parameter	Preferred condition
Lithology	Sandstones preferred
Wettability	Water-wet
Current oil saturation	Above residual oil saturation
Porosity type (matrix/fractures)	Matrix preferred
Gas cap	See comments
Aquifer	Edge aquifer tolerated
Salinity/hardness	See comments
Dykstra-Parsons and facies variations	$0.1 < DP < 0.8$
Clays	Low (see comments)
Water-cut	See comments
Flooding pattern and spacing	Confined – small spacing

3.3.2.1. Lithology

Review of historical CEOR projects finds that more floods have been performed in sandstones compared to carbonates. For instance, between 1971 and 1990, 320 pilot projects or full field floods were referenced, among which only 57 were conducted in carbonate reservoirs [20]. A possible explanation behind these statistics lies in the types of polymers commercially available at that time, their different advantages, and the relative simplicity of sandstone reservoirs. For instance, synthetic anionic polyacrylamides display a high viscosifying power and very high molecular weights, and they are cost-effective to produce. Their anionic nature minimizes possible ionic interactions with minerals in siliciclastic reservoirs, which mainly display negative charges at reservoir pH. Similarly, clays are always deficient in positive charges due to cation substitution (Si^{4+} replaced by Al^{3+} in tetrahedral sheets, and Al^{3+} replaced by Fe^{2+} or Mg^{2+} in octahedral sheets, for instance). Therefore, interlayer cations between the platelets are required to balance the face negative charges, allowing in some cases bonding with the injected anionic chemicals and irreversible loss by adsorption.

For carbonates, injecting anionic polymers with adapted chemistry is possible but requires extensive laboratory studies to minimize excessive retention or, in some cases, precipitation. Cationic polymers have been considered, but they are more expensive to produce compared to anionic polymers, are highly shear sensitive, and display lower molecular weights on average, rendering the whole process more complex.

3.3.2.2. Wettability

Before recent studies on polymer viscoelasticity – which are ongoing – polymers were only considered to displace mobile oil; decreasing the residual oil saturation required the use of surface active agents (surfactants) in combination with polymers and/or alkaline agents. Therefore, only water-wet reservoirs have been considered for straight polymer flooding applications, which can essentially be narrowed down to sandstone reservoirs with small amounts of clays. For carbonates (which are often oil-wet), wettability is more of an issue and can compromise the sole injection of polymers (Figure 3.3). The addition of surfactants is often required to mobilize the trapped oil and render the

---

Figure 3.3

Comparison of fluid distribution in a water-wet and an oil-wet reservoir

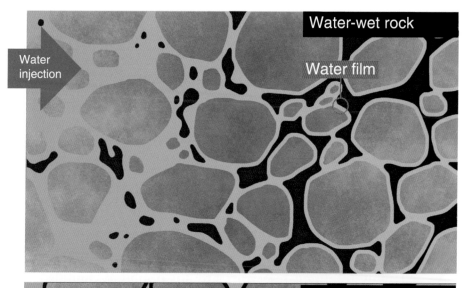

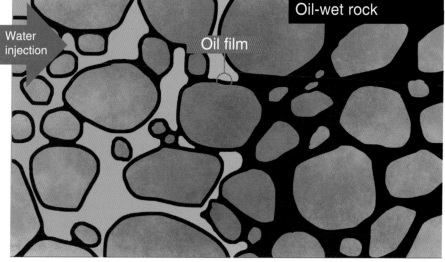

process viable. Assessing wettability is therefore critical before starting any CEOR screening process.

3.3.2.3. Current Oil Saturation

If we assume, for the sake of simplicity, that polymers can only push mobile oil, it is necessary to carefully assess the current oil saturation in the zones of interest to make sure polymer flooding can recover substantial volumes of hydrocarbons. A current saturation above the residual oil saturation is therefore necessary to make the process viable technically. For economic purposes, the higher the oil saturation, the higher the potential revenue, thereby further supporting the early adoption of EOR techniques.

The oil saturation can be assessed by analyzing production history and calculating recovery factors or by performing chemical tracer tests. This information may also be gleaned from the logs of in-fill wells drilled later in development, which may indicate previous flood performance and the vertical displacement profile in the interwell regions.

3.3.2.4. Porosity Type

Highly fractured reservoirs are not optimum candidates for polymer flooding, given the probability of early polymer breakthrough and overall poor conformance. In this case, gel treatments with polymer flooding are possible but require strong reservoir understanding.

3.3.2.5. Gas Cap

The presence of a gas cap is not necessarily a limiting factor, depending on reservoir configuration, well placement, and production history. Obviously, the main problem with the presence of gas is compressibility: a gas cap is good at stealing pressure and can delay the efficiency of any CEOR or pressure-support method.

3.3.2.6. Aquifer

The presence of a bottom aquifer can be detrimental to polymer flooding if injection occurs below the water/oil contact. Dilution of chemicals should obviously be avoided; and, sometimes, well workover and recompletion are necessary to better isolate the injection target from the aquifer. In vertical wells, for instance, cementing the bottom of the well and re-perforating the top of the oil-bearing layer can be an option. In some reservoir configurations, the presence of an edge aquifer is not necessarily limiting. This depends on field configuration, well placement, and production history. For instance, consider an anticline reservoir with an edge aquifer, peripheral injection, and production at the top of the structure. In this case, it is possible to start by injecting a high-viscosity slug to stop the aquifer from moving upward, while also using gravity underride and viscosity difference as tools to maximize sweep efficiency.

3.3.2.7. Salinity/Hardness

Polymer viscosity is impacted by the salt content of the injection water and especially the divalent content (calcium, magnesium). These cations will screen the negative charges of functional groups on the polymer molecules, leading to a hydrodynamic volume decrease. The greater the hardness, the higher the polymer concentration required to reach a given viscosity (Figure 3.4). The choice of the most appropriate chemistry is dictated by considering both the salinity and the temperature. For instance, a standard copolymer of acrylamide and acrylic acid can sustain temperatures above 100 °C if no divalent is present in the injection water [21] (as in alkali-surfactant-polymer [ASP] processes, for example). Conversely, adding a small percentage of sulfonated monomer might be required to maintain good stability in a brine with $200\,\mathrm{g\,l^{-1}}$ total dissolved salts (TDS) with 1% of divalent in a reservoir at 30 °C. Overall,

Figure 3.4 Impact of increasing salinity (and R⁺) on the viscosity of a copolymer of acrylamide and acrylic acid (Flopaam 3630S)

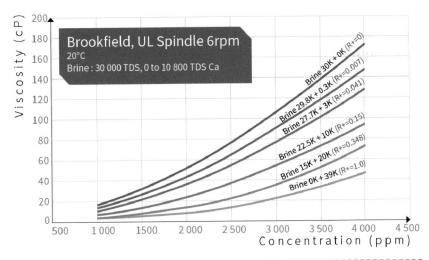

salinity will mainly impact economics, since improvements in chemistry have led to the design of polymers that are stable even in harsh conditions.

3.3.2.8. Dykstra-Parsons

The Dykstra-Parsons coefficient characterizes permeability variations and is a common descriptor of reservoir heterogeneity. Its value ranges from 0 to 1, with 0 being a homogeneous reservoir and 1 a very heterogeneous reservoir. The goal of polymer injection is to improve reservoir areal and volumetric sweep efficiencies; the value of such a process is therefore totally justified when the reservoir has some degree of heterogeneity. However, if the Dykstra-Parsons coefficient approaches 1, and the permeability contrast is too great, there is a high probability of early chemical breakthrough, especially if the viscosity injected is too low.

Ranges for Dykstra-Parsons values can be set between 0.1 and 0.8 for polymer flood applicability, keeping in mind that no reservoir is perfectly homogenous and a reservoir with a value greater than 0.8 will likely require conformance treatments before polymer flooding to avoid premature polymer break-through. The ratio of vertical permeability to horizontal per-meability is also important in order to select the best injection strategy and use gravity to maximize the overall efficiency. This is especially important to consider for reservoirs with large net pay and well spacings.

3.3.2.9. Clays

The presence of clays can be detrimental to polymer (or any chemical) propagation. The high surface area of many clay fam-ilies (especially the smectite type) favors irreversible polymer adsorption, delaying the viscous front and subsequently the oil bank, and inevitably decreasing the overall efficiency. It is possible to quantify the retention with core flooding tests using reservoir cores, possibly from different reservoir locations in order to be representative of the various geological facies. However, even if a value is obtained, the lack of complete reservoir knowledge and facies distribution will render this value almost obsolete, or opti-mistic at best. But even this optimistic retention value can be used to give an idea of how much additional polymer may be required to achieve the target pore volume injected. If retention is expected because of high clay content, several options can be considered:

- Inject a chemistry that is less prone to adsorption, such as polymers containing sulfonated groups (ATBS) [22, 23].

- Inject a sacrificial agent [24]. However, this strategy often requires the use of an extra chemical, which, in the end, has to be injected over the same reservoir pore volume (or, more precisely, mass, since it is a mass balance) as the CEOR cocktail to be efficient at reducing adsorption. This can have a substantial extra cost in addition to other possible design considerations, such as how to ensure that the sacri-ficial agent will contact the same zones as the CEOR cocktail if the viscosity is different.

- Slightly overdose the injected polymer, to compensate for the adsorption calculated in the laboratory. This strategy is probably the simplest and most economical since it does not require any additional chemicals or equipment. For example, depending on the adsorption value obtained through laboratory tests, the injection strategy could consist of injected 1050 ppm active polymer for a target at 1000 ppm to balance retention. Alternatively, a brief higher-concentration/viscosity slug may be injected first, when injectivity is the best, before switching back to the target, working-viscosity slug.

3.3.2.10. Water-cut

Concerns are sometimes raised about the efficiency of CEOR methods in zones with very high water-cut (>95%). From a practical standpoint, a high water-cut in the producers has three main possible origins:

- Almost all of the hydrocarbons in place have been recovered, the field will keep producing water, and the economic limit has been reached. In that case, there is little interest in evaluating EOR methods. The field can be abandoned. Obviously, this is rarely the case.

- Water injection has been inefficient at displacing the hydrocarbon because of a difference in mobility between the fluids and/or large-scale heterogeneities. In that case, considering CEOR methods makes sense to improve the recovery factor.

- Production wells are drawing water from an exterior source such as an aquifer, through fractures, coning, or casing leaks. In that case, identifying and treating the problem is necessary before screening EOR methods.

When substantial volumes of oil are left in the reservoir and the water-cut is high because of poor sweep efficiency, polymer flooding will increase the recovery factor. A drawback, depending on the injection history, reservoir configuration, and water saturation, might be the delay needed before observing changes in water-cut and oil cuts, since the polymer slug will also displace the water present in the reservoir. This is another reason EOR methods should be started as early as possible, to mitigate early onset of two-phase production.

3.3.2.11. Flooding Pattern and Spacing

There are usually two stages in EOR deployment. The first consists of collecting and evaluating fluid-fluid and fluid-rock, and other laboratory screening to obtain parameters for further consideration. The second includes validating the technology through pilot tests, and gathering all kinds of relevant data to decide about further expansion to a full-field recovery scheme or cessation. In both cases, significant lessons learned should be applied to improve the process application. The aim of a pilot is to prove the viability of a given technology as quickly as possible. It is therefore compulsory to have a confined pattern of close well spacing and relatively small net pay thickness to (i) isolate oil production from EOR, (ii) minimize uncertainty and chemical losses, and (iii) minimize the response time.

3.4. Conclusions

If the first three criteria are met (siliciclastic, water-wet reservoir with high remaining, mobile oil saturation), there is a high probability that polymer injection (alone) will be technically viable. The injection strategy (development, rates) and fine-tuning of the chemical selection will dictate the economic viability of the process. These aspects will be discussed in more detail later in this book.

In a Nutshell

The reservoir applicability of the polymer flooding technique has been greatly extended over the last 20 years. It applies now to difficult reservoirs with high temperature and salinity, and it extends the productive life of more viscous oils previously not considered for flooding operations. Simple screening criteria can be used to select candidate reservoirs where the technology can be technically viable. The next step is to design the best polymer chemical system and injection strategy to make the project economically viable.

References

[1] Thomas, A. (2016). Polymer flooding. In: *Chemical Enhanced Oil Recovery (cEOR) - a Practical Overview* (ed. L. Romero-Zerón). InTech https://doi.org/10.5772/64623.

[2] Muskat, M. (1949). *Physical Principles of Oil Production.* New-York City: McGraw-Hill Book Co.

[3] Stiles, W.E. (1949). Use of permeability distribution in waterflood calculations. *Journal of Petroleum Technology* 1: 9–13.

[4] Dykstra, H. and Parsons, R.L. (1950). *The Prediction of Oil Recovery by Waterflood. Secondary Recovery of Oil in the United States, Principles and Practices*, 2e, 160–174. American Petroleum Institute.

[5] Aronofsky, J.S. and Ramney, H.J. Jr. (1956). Mobility ratio-its influence on injection or production histories in five-spot waterflood. *Journal of Petroleum Technology* 8: 205–210.

[6] Dyes, A.B., Caudle, B.H., and Erickson, R.A. (1954). Oil production after breakthrough as influenced by mobility ratio. *Journal of Petroleum Technology* 6: 27–32.

[7] Pye, D.J. (1964). Improved secondary recovery by control of water mobility. *Journal of Petroleum Technology* 16: 911–916. AIME, 231.

[8] Sandiford, B.B. (1964). Laboratory and field studies of waterflood using polymer solutions to increase oil recoveries. *Journal of Petroleum Technology* 16: 917–922, AIME, 211.

[9] Leverett, M.C. (1941). Capillary behavior in porous solids. *Transactions of the AIME* 142: 152–169.

[10] Buckley, S.E. and Leverett, M.C. (1942). Mechanism of fluid displacement in sands. *Transactions of the AIME* 146: 107–116.

[11] Welge, H.J. (1952). A simplified method for computing oil recovery by gas or water drive. *Journal of Petroleum Technology* 4: 91.

[12] Craig, F.C. Jr. (1971). *The Reservoir Engineering Aspects of Waterflooding*, Monograph Series, SPE, vol. 3, 35–38. Richardson, TX: Society of Petroleum Engineering.

[13] Willhite, G.P. (1986). *Waterflooding*, SPE Textbook Series, vol. 3, 64–67. Richardson, TX: Society of Petroleum Engineers.

[14] Sorbie, K.S. (1991). *Polymer-Improved Oil Recovery.* Boca Raton, FL: CRC Press, Inc.

[15] Chang, H. L. (1978). Polymer flooding technology: yesterday, today and tomorrow. Paper SPE7043 presented at the Fifth Symposium on Improved Methods for Oil Recovery, Tulsa, OK, 16–19 April. https://doi.org/10.2118/7043-PA.

[16] Seright, R.S. (2010). Potential for polymer flooding viscous oils. *SPE Reservoir Evaluation and Engineering* 13 (6): 730–740.

[17] Vermolen, E.C.M., Van Haasterecht M.J.T., Masalmeh S.K., et al. (2011). Pushing the envelope for polymer flooding towards high-temperature and high-salinity reservoirs with polyacrylamide based ter-polymers. Paper SPE 141497 presented at SPE Middle East Oil and Gas Show and Conference. Manama, Bahrain, 25–28 September. https://doi.org/10.2118/141497-MS.

[18] Gaillard N., Giovannetti B., and Favero C. (2010). Improved oil recovery using thermally and chemically stable compositions based on co and ter-polymers containing acrylamide. Paper SPE 129756 presented at SPE Improved Oil Recovery Symposium, Tulsa, Oklahoma, USA, 24–28 April. https://doi.org/10.2118/129756-MS.

[19] Abirov Z., Abirov R., Mazbayev Y., et al. (2015). Case study of successful pilot polymer flooding in the south Turgay Basin's oilfield. Paper SPE 177339 presented at the SPE Annual Caspian Technical Conference and Exhibition, Baku, Azerbaijan, 4–6 November. https://doi.org/10.2118/177339-MS.

[20] Office of Technology Assessment. 1978.Enhanced Oil Recovery Potential in the United States. Library of Congress Catalog Card Number 77-600063.

[21] Seright, R.S., Campbell, A.R., Mozley, P.S., and Han, P. (2010). Stability of partially hydrolyzed polyacrylamides at elevated temperatures in the absence of divalent cations. *SPE Journal* 15: 341–348.

[22] Hollander, A.F., Somasundaran, P., and Gryte, C.C. (1981). Adsorption characteristics of polyacrylamide and sulfonate-containing polyacrylamide copolymers on sodium kaolinite. *Journal of Applied Polymer Science* 26: 2123–2138.

[23] Rashidi, M., Blokhus, A.M., and Skauge, A. (2010). Viscosity and retention of sulfonated polyacrylamide polymers at high temperature. *Journal of Applied Polymer Science* https://doi.org/10.1002/app.33056.

[24] Shamsijazeyi H., Hirasaki G., and Verduzco R. (2013). Sacrificial agent for reducing adsorption of anionic surfactants. Paper SPE 164061 presented at the SPE International Symposium on Oilfield Chemistry, The Woodlands, Texas, USA, 8–10 April. https://doi.org/10.2118/164061-MS.

Chapter four:

Polymers

Polyacrylamide chemistry, manufacturing, and rheological properties will
be discussed in this chapter. Typical
qualification tests and corresponding
laboratory procedures will be detailed,
providing general selection guidelines
for polymer flooding design.

Essentials of Polymer Flooding Technique, First Edition. Antoine Thomas.
© 2019 John Wiley & Sons Ltd. Published 2019 by John Wiley & Sons Ltd.

4.1. Introduction

Polymers are macromolecular chemical compounds that exhibit several specific features:

- Their molecular weight is high; it can range from a few thousand to several million daltons. In enhanced oil recovery (EOR) processes, average molecular weights can range between 3 and 35 million daltons.

- They are made up of repeat units named after the monomer used. Polyacrylamide, for example, is a polymer chain that consists of repeating units originating from acrylamide (Figure 4.1).

- Polymer synthesis is a complex process that generally involves several steps, each of them resulting in the formation of a bond and chain extension.

- Polymers are usually polydisperse, meaning macromolecules have the same chemical composition but different chain lengths coexisting within a sample.

In the oil and gas industry, two families are usually considered when attempting to increase the viscosity of water for reservoir sweep improvement: biopolymers (polysaccharides) and

----

Figure 4.1

From monomers to polymers: the polymerization process

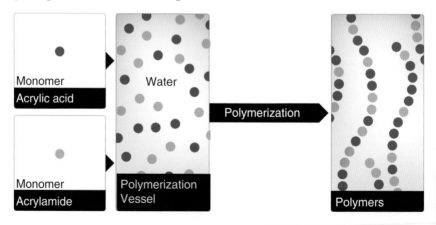

Monomer
Acrylic acid

Water

Polymerization

Monomer
Acrylamide

Polymerization
Vessel

Polymers

synthetic polymers. Only synthetic polymers and particularly polyacrylamides will be described hereafter.

4.2. Polyacrylamide – Generalities

4.2.1. Introduction

Polyacrylamide polymers and their derivatives are made by the polymerization of acrylamide alone or with other monomers to impart positive or negative charges on the molecule backbone. They have many uses, the largest being as a dewatering aid in municipal and industrial wastewater treatment plants. Other uses include flocculants in the mining industry, viscosifiers and drag reducers in oil and gas, papermaking aids, soil stabilizers, additives for textiles, paints, thickeners in cosmetics, and many more.

4.2.2. Monomers

In addition to acrylamide, several other monomers can be incorporated during the synthesis to fit specific purposes. In chemical EOR applications, the vast majority of projects were/are implemented in sandstone reservoirs using anionic polymers to minimize adsorption. Since acrylamide is nonionic, a negative charge can be brought either by incorporating a specific monomer such as acrylic acid or by hydrolyzing the amide moieties with caustic soda, yielding carboxylic acid function groups. The first process is called *copolymerization*, and the second is referred to as post-hydrolysis (see Section 4.2.3.1).

4.2.2.1. Acrylamide

Acrylamide is a white, crystalline, water-soluble compound derived from acrylonitrile [1]. It contains an electron-deficient double bond and an amide group and undergoes chemical reactions typical of these two functionalities [2] (Figure 4.2). Acrylonitrile is produced by reacting propylene, ammonia, and oxygen in a single fluidized bed of complex catalyst

Figure 4.2 | # Acrylamide monomer

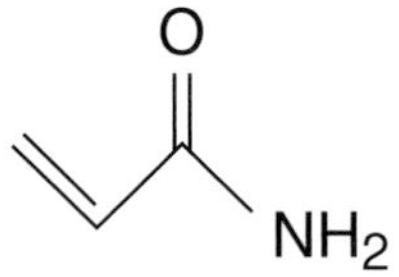

(Sohio process, first operated commercially in 1960) [3]. Th transformation process of acrylonitrile to acrylamide can b achieved with the use of a reduced copper catalyst but results i a poor yield. It is possible to improve the reaction by using ar enzyme (nitrile hydratase) originally patented in 1973 b ENSAM (Ecole Nationale Supérieure Agronomique d Montpellier) after joint development with SNF. The patent wa then licensed to Nitto Chemical Industry and commercialized i 1985, with continuous improvements of the microorganism eve since.

Since 1971, acrylamide has been preferentially handled in 50% active aqueous solutions to eliminate solids handling and decreas costs. Solutions with stabilizers (cupric ions, ethylenediaminetet raacetic acid [EDTA], etc.) are kept in coated tanks and store away from sunlight to prevent heating and polymerization Acrylamide is a hazardous chemical classified as probably carci nogenic to humans (International Agency for Research o Cancer – Group 2A). Recently, the European Food Safet Authority (EFSA) issued a note saying that "results from huma studies provide limited and inconsistent evidence of increase risk of developing cancer (of the kidney, endometrium an ovaries) in association with dietary exposure to acrylamide." I added that "studies on workers exposed to acrylamide in th workplace show an increased risk of disorders to the nervou system" (http://www.efsa.europa.eu/en/search/site/acrylamide) In food, acrylamide appears during the Maillard reaction b heating (>120 °C), and some recent studies have shown compel ling support for the hypothesis of a sustained endogenous acryl amide formation in the human body [4]. In contrast, polymers o acrylamide are known to exhibit a very low toxicity [5].

4.2.2.2. Acrylic Acid

The most common anionic polyacrylamides used in the EOR industry are copolymers of acrylamide and acrylic acid (Figure 4.3). Acrylic acid is a moderately strong carboxylic acid; it was first prepared in 1847 by air oxidation of acrolein [2, 6]. Today, the principal demand for acrylic acid remains for super-absorbents for use in hygienic products, including diapers.

Acrylic acid monomer

Figure 4.3

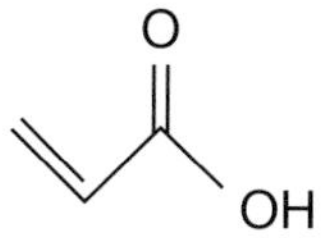

4.2.2.3. ATBS

ATBS (or AMPS™) stands for acrylamido tertiary butyl sulfonic acid (or, strictly, acrylamido-2-methylpropane sulfonic acid). ATBS is made by the Ritter reaction of acrylonitrile and isobutylene in the presence of sulfuric acid and water (Figure 4.4). This monomer was studied to overcome the stability issues of amide groups at high temperature. The dimethyl and sulfomethyl groups sterically hinder the amide function and provide thermal and hydrolytic stability to ATBS-containing polymers. Excessive hydrolysis of amide groups to carboxylate is a major cause of instability of polyacrylamides; it can cause polymer precipitation and, therefore, viscosity loss.

ATBS monomer

Figure 4.4

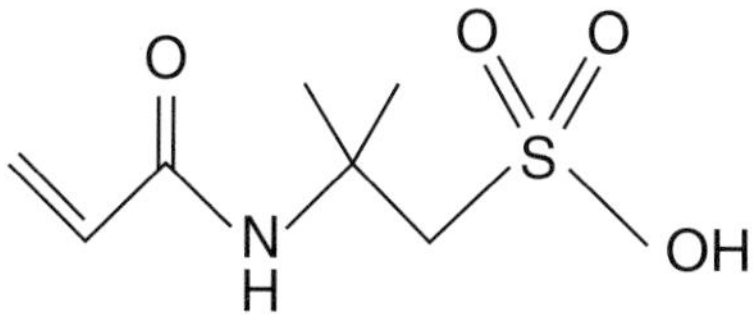

N-Vinylpyrrolidone (NVP) is an organic compound produced from the reaction of acetylene with 2-pyrrolidone. This monomer can protect the neighboring acrylamide groups from hydrolysis and improves temperature stability [7–12]. However, the reactivity of this monomer is relatively low, resulting in composition drifts and low molecular weights.

It is possible to add other monomers to introduce additional features, such as the creation of associative polymers, for instance. However, the precise chemical composition will not be discussed here.

The selection of the most appropriate chemistry, monomer ratios, and molecular weights depends on reservoir characteristics. These aspects will be discussed later in the book.

4.2.3. Polymerization Processes

Acrylamide can be polymerized via free radical initiation or by an anionic mechanism [2, 6, 13]. The free radical polymerization uses substances that generate radicals by homolytic scission when heated or irradiated, for instance. The most common free radical initiators are peroxides, hydroperoxides, peresters, and aliphatic azo compounds such as AIBN (azobisisobutyronitrile).

The mechanism of free radical polymerization is divided, simplistically, into three steps: initiation, propagation, and termination. The initiation reaction is the attack of a monomer by a radical originating from the initiator. This reaction is then repeated thousands of times for each newly formed chain (propagation). Termination is enabled by the combination of two propagating radicals or by the addition of a transfer agent involving mutual destruction of radicals and impacting the final molecular weight distribution of the product [2]. This is an exothermic reaction, conducted in absence of oxygen, which allows building very high-molecular-weight polymers thanks to the high conversion rates (Figure 4.5).

Radical polymerization

Figure 4.5

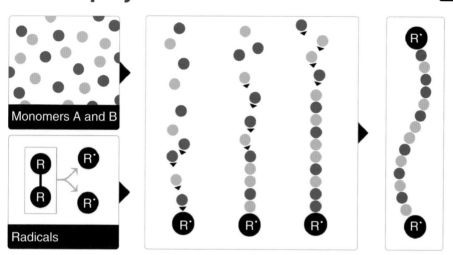

There are several manufacturing paths to make polyacrylamides, including the gel polymerization, dry bead, microemulsion, and inverse emulsion polymerization processes. The most economical means of production is the gel process.

4.2.3.1. Gel Polymerization Process

An aqueous solution containing the monomers, initiators, and other additives is polymerized under adiabatic conditions. At the end of this exothermic reaction, a gel is formed, which is crushed in small pieces and dried to finally obtain a powder (Figure 4.6).

For EOR applications where polymers with specific properties and ionicity are required, two processes are generally used to obtain the most appropriate molecules:

- *Copolymerization process.* In this case, acrylamide is copolymerized with other monomers such as acrylic acid to obtain an anionic copolymer. The charge repartition occurs relatively homogeneously along the polymer backbone.

| Figure 4.6 | **Gel polymerization process** |

▲ Gel obtained at the end of polymerization

▲ Powder obtained afer grinding, drying, and sieving

- *Post-hydrolysis process.* Acrylamide is polymerized alone during this process. In the end, a homopolymer of acrylamide is obtained, with a very high molecular weight. After crushing, caustic soda is added over the grains to hydrolyze a portion of the acrylamide moieties to impart the anionicity. In this case, the charge repartition is more random than for the copolymerization process.

The final product is a white granular powder with an average active content above 85% and grain size below 1 μm. In the field, equipment is required to dissolve and hydrate the polymer. This will be detailed later.

4.2.3.2. Inverse Emulsion Polymerization Process

Polyacrylamides in inverse emulsions are a dispersion of internal aqueous polymer particles in an organic solvent external phase (oils) [14]. The first step consists of placing water with the chemical reagents, monomers, and oil in a polymerization vessel. A careful selection of phase ratio and surfactants

Inverse emulsion preparation. Emulsification steps: mixing of oil and water phase before polymerization

Figure 4.7

(emulsifier) is needed to perform the emulsification. Polymerization occurs inside the small water droplets, resulting in a stable dispersion of polymer particles with an average size ranging from 200 nm to 1 μm (Figure 4.7).

The final product has an active polymer content between 25% and 40%. It is possible to reach higher levels (up to 60 wt%) using a distillation process. In the field, an inversion process is required to move from a water-in-oil emulsion to an oil-in-water emulsion to allow polymer uncoiling and viscosity buildup. This step requires a careful design of the surfactant package in the emulsion and the equipment used to invert the product.

4.3. Polymer Selection Guidelines

4.3.1. Generalities

The choice of the most appropriate polymer form and chemistry depends on a few factors that will be discussed in the following paragraphs. In order to preselect several polymer candidates, it is necessary to know the reservoir temperature, the average permeability, and the salinity of the injection water.

Reservoir temperature and the salinity of injection water are required to optimize the chemistry. The permeability is needed to adapt the average molecular weight of the final product to ensure a smooth propagation of the molecules through the rock.

4.3.1.1. Polymer Form

The choice of the product form depends mainly on field constraints, location, and logistics. For chemical EOR projects, two product forms are usually considered: powder and inverse emulsion (Figure 4.8). The liquid form is usually preferred offshore when the local weather conditions, footprint, and logistics limit the use or transfer of powders. From an economic standpoint, powders are more cost-efficient than emulsions but also require a larger footprint since the dissolution and hydration equipment is larger. Engineering and logistical studies are required to choose the most favorable option, depending on the existing infrastructures or other field constraints.

Figure 4.8 **Powder vs. emulsion product forms.**

4.3.1.2. Polymer Chemistry

Selecting the optimized monomeric composition is paramount to ensure that the target viscosity will be maintained throughout the journey in the reservoir. Above 60 °C (but also depending on pH and time), the acrylamide groups along the polymer backbone experience hydrolysis and form acrylate groups. If there is a significant concentration of divalent cations in the brine, the polymer can precipitate out, leading to an irreversible viscosity drop. Usually, copolymers of acrylamide and acrylic acid are considered up to 75 °C in mild brines. However, in the absence of oxygen and divalent cations, such polymers can remain stable up to 120 °C [15].

At higher temperatures and/or when the divalent content increases, it is possible to incorporate ATBS to prevent precipitation and extend the temperature range of application. Numerous studies have been performed to evaluate the benefits of incorporating this sulfonated monomer; the main result is that ATBS increases salinity and temperature tolerance up to 95 °C [7–10, 16–24]. Additional studies have shown that the presence of ATBS increases shear stability and decreases retention in reservoirs [18, 22, 25–29].

For higher temperatures, it is possible to add NVP to protect acrylamide from hydrolysis, as discussed by several authors [7–10]. The temperature limit for such polymers can be extended up to 140 °C in certain conditions. However, these polymers have lower molecular weights than conventional copolymers of acrylamide and acrylic acid, for instance, because of reactivity issues during polymerization. This often requires a higher polymer dosage to reach a given viscosity, thus leading to higher costs.

4.3.1.3. Polymer Molecular Weight

Common polymers used in EOR have molecular weights between 2 and 35 million g (g mol)$^{-1}$ or daltons. The final molecular weight distribution can be controlled during the polymerization step by the transfer agent present in the polymerization vessel. Polyacrylamides are characterized by a polydispersity index (PDI), meaning that every sample is characterized by a molecular weight distribution following a Gaussian-type curve [29, 30] (Figure 4.9). Research work is ongoing to develop techniques to measure the molecular weight distribution more precisely, since conventional gel permeation chromatography (GPC) techniques are limited for such high molecular weights. The range of molecular weights is usually determined by intrinsic viscosity measurements.

Figure 4.9 **Polymer molecular weight distribution.**

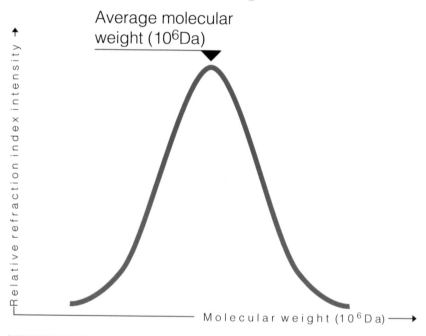

4.3.2. Polymer Selection

General guidelines will be provided here based on field cases and other studies. It should help narrow the choice of candidates to start laboratory studies. Three parameters are needed to start: reservoir temperature, salinity (and, more precisely, a parameter called R^+), and permeability.

The parameter R^+ is defined as:

$$R^+ = \frac{[C_{cat\ div}]}{[C_{cat\ mono}] + [C_{cat\ div}]}$$

where $[C_{cat\ div}]$ is the number of moles of divalent cations in the brine and $[C_{cat\ mono}]$ is the number of moles of monovalent cations in the brine.

Gaillard et al. [31] compared the salinity and shear tolerance of different chemistries. Standard copolymers are efficient viscosifiers in brine with R^+ below 0.05 and total salinity below 50 000 ppm. Terpolymers with ATBS are needed when R^+ is above 0.05 with total salinity below 100 000 ppm. When R^+ becomes greater than 0.1 and salinity is below 100 000 ppm, copolymers of acrylamide and ATBS are required. It was also shown that copolymers containing ATBS are more shear resistant than any other type of chemistry.

4.3.2.1. Molecular Weight

For EOR, knowing the average reservoir permeability and comparing with existing laboratory studies and field cases helps indicate the best average molecular weight for the chosen, allowing good propagation in the matrix (Table 4.1).

This table should be used cautiously, since chemistry and salinity can greatly impact the hydrodynamic volume of the polymer, sometimes favoring its propagation in lower permeability. Some injection strategies also consist of pre-shearing high-molecular-weight polymers to enhance injectivity.

4.3.3. Other Polymer Families

4.3.3.1. Associative Polymers

A major concern in polymer flooding is the cost-effectiveness of the chemical and its ability to provide a sufficient resistance

Table 4.1 **Empiric correlation between average molecular weight and absolute rock permeability.**

Average molecular weight (million Da)	Minimum permeability (mD)
>20	>1000
18–20	>750
15–18	>500
12–15	>350
8–12	>200
5–8	>100
1–5	>10

factor (effective viscosity of polymer in porous media relative to water) in the reservoir to displace oil. Alternatives to regular partially hydrolyzed polyacrylamide (HPAM) have been studied since the 1980s, including associative polymers [32–41]. For these molecules, a small fraction of a hydrophobic monomer is incorporated in the polymer to promote intermolecular associations and enable the construction of a reversible 3D network that can perpetually form and break, thereby increasing the resistance factor in the flowing fluid.

Several studies have been performed, highlighting concerns about the propagation of associative polymers in porous media [42]. A first possible explanation for mediocre propagation can be linked to the complexes formed by the associative polymers, which can yield very high resistance factors at low flow rates and, therefore, greatly impact the flow of the solution in the reservoir. The second is polymer retention, which is much higher for such chemistries compared to regular HPAM. The changes in viscosity are also quite abrupt and may impact the efficiency of oil displacement inside the reservoir if adsorption occurs. Finally, given the amphiphilic nature of the molecule, interactions with the oil are likely, and the formation of very stable emulsions at the production side is possible. A careful laboratory design is required, including long cores with multiple pressure taps and a precise quantification of retention.

4.3.3.2. Thermoresponsive Polymers

For standard co- or terpolymers, viscosity decreases with temperature. Given this characteristic, maintaining efficient mobility control in hot reservoirs can become difficult. Alternative chemistries with the ability to associate under specific conditions have been developed and studied for various applications [43]. As with thermoresponsive polymers, the concept involves water-soluble chains with blocks or side groups with lower critical solution temperature (LCST) moieties, allowing associations when that specific temperature is reached [44].

For EOR or conformance applications, using polymers whose viscosity increases with temperature can have several advantages. The first is injecting a solution with a low viscosity, favoring short-term injectivity. A second is being able to maintain reasonable mobility control even at high temperatures and extend the envelope of applications of polymer flooding to hot reservoirs with very saline injection water. A typical viscosity profile as a function of temperature is given in Figure 4.10.

As with regular polymers, several laboratory tests, including viscosity profiles and core flooding, are required to assess the changes in resistance factor in a porous medium. With this type of chemistry (and any type of polymer with associations), a viscosity measured with a rheometer will be much different than the in situ viscosity or resistance factor of the flow fluid in a piece of rock.

4.4. Polymer Characteristics and Rheology

In this section, we will review the main characteristics of polyacrylamides in solution.

Figure 4.10 # Viscosity vs. temperature for thermoresponsive polymers.

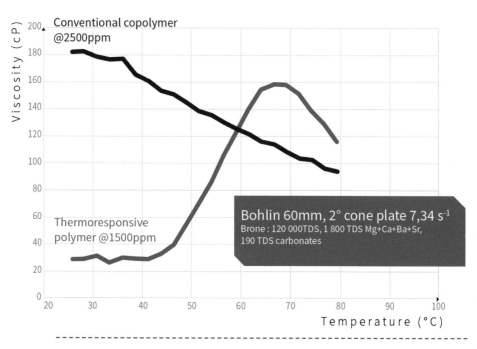

4.4.1. Viscosity

4.4.1.1. Generalities

Viscosity is one of the most frequently applied methods to characterize polymer solutions because of the ease and rapidity of measurement. The addition of macromolecules with a very large size in a solvent will allow measuring a viscosity enhancement over the pure solvent. A polymer is able to increase the viscosity of a medium only if the polymer chains are deployed. This means the interactions between polymer/solvent should be more favorable from an energetic standpoint than the polymer/polymer interactions. Also, in a polar medium like water, an anionic polymer will be able to expand due to electrostatic repulsions. The larger the hydrodynamic volume, the higher the resulting viscosity (Figure 4.11).

Solvent viscosification process by polymer addition.

Figure 4.11

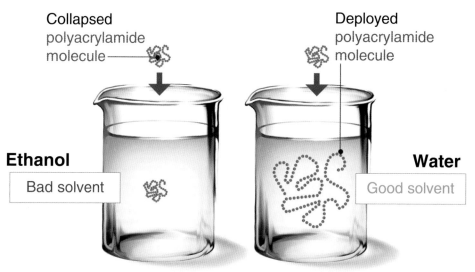

The *specific viscosity* expresses the degree to which the viscosity of a solvent is enhanced:

$$\eta_{sp} = \frac{(\eta - \eta_s)}{\eta_s}$$

where η and η_s are the solution and solvent viscosities, respectively.

The *intrinsic viscosity* is used to determine the average molecular weight; it describes the contribution of an isolated polymer to the viscosity of the solution in the absence of interpolymer interactions [45].

The factors affecting viscosity are as follows:

- Molecular weight
- Temperature
- Shear rate
- Chain stiffness

- Topology
- Solvent quality (solvent and its composition, e.g. water and salt content)
- Polydispersity

Figure 4.12 illustrates the impact of salt content (via R^+) on the viscosity of acrylamide and acrylic acid copolymers. As mentioned earlier, the addition of salts will screen the negative charges on the polymer backbones and impact the electrostatic repulsions at the origin of polymer deployment; a lower hydrodynamic volume will result, leading to a viscosity drop.

Figure 4.12 **Viscosity variations vs. R^+ for polymers with increasing percentages of ATBS (from 3630S to AN125VHM).**

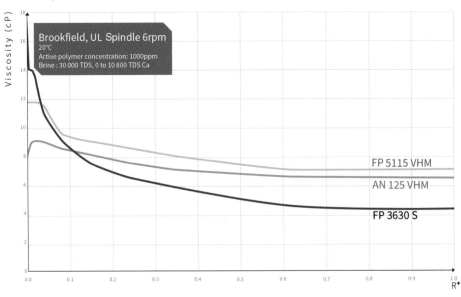

For regular polymers, viscosity decreases as temperature increases. At a fixed temperature and letting a polymer solution age, the hydrolysis of acrylamide to acrylates moieties increases the percentage of negative charges on the backbone. If no divalent cations are present in the solvent, the electrostatic repulsions increase, and so does viscosity. If divalent cations are present, the screening effect can lead to polymer precipitation and, subsequently, a decrease in viscosity. The devices and procedures used to determine viscosity will be discussed in a separate section. In summary, the viscosifying capability of HPAM is linked to the level of entanglement and the intra- and intermolecular electrostatic repulsions between polymer macromolecules.

4.4.2. Rheology

Polyacrylamide solutions are non-Newtonian fluids, meaning viscosity is dependent upon the shear rate applied. Specifically, such polymer solutions show a pseudoplastic behavior: viscosity decreases as the shear rate increases (also called *shear-thinning*). Such behavior can be described using a power-law model, as shown in Figure 4.13 [46]. This process is reversible: up to a certain value, if shear is stopped, viscosity returns to its original value.

The viscosity versus shear rate profile is dependent upon several factors including molecular weight, chemistry, salinity, etc. Also, the rheology in porous media is quite different than what is observed with rheometers, due to viscoelasticity and the

----

Apparent viscosity of a polyacrylamide solution as a function of shear rate.

Figure 4.13

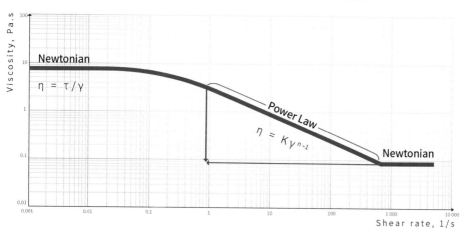

dynamics of polymer conformation in solution. These complex topics will not be addressed in detail in this book.

4.4.3. Solubility

When considering polyacrylamides in powder form, it is important to consider the solubility of the product to ensure that no insoluble or swollen particles remain before injection. So-called *fish eyes* (insoluble particles) can be detrimental to injectivity and impair the efficiency of the EOR technique. Filtration tests are usually used to ensure that the polymer batch corresponds to the established specifications given in the certificate of analysis.

4.5. Polymer Stability

Polyacrylamide macromolecules are sensitive to several types of degradations that can occur at the injection or production side in EOR processes. It is paramount to protect the polymer at the injection side to ensure that the right viscosity is injected and integrity is sustained while flowing in the reservoir to displace oil. Obviously, strategy for polymer injection and protection encompasses a strict and well-designed quality control program, the implementation of which should also minimize uncertainties linked to sampling and viscosity measurements.

At the production side, polymer degradation is less of an issue. Degradation at that point in the process can help minimize the potential impact of viscosity on the separation of oil and water at the treatment facilities. Only what happens at the injection side will be discussed in this section.

4.5.1. Chemical Degradation

Chemical or oxidative degradation is, in the vast majority of cases, a free radical chain reaction that can be greatly accelerated upon irradiation by UV, for example. A radical site can be created on the polymer backbone by removing a tertiary hydrogen atom. A peroxy radical is generated when this site binds to an oxygen molecule. The latter can in turn take away another tertiary hydrogen, or it can induce chain scission [2]. In EOR processes, such free radicals are usually generated by oxidation-reduction reactions between oxygen dissolved in the water and either iron (II) or H_2S (hydrogen sulfide). Chain scission will lead to a reduction of polymer hydrodynamic volume, resulting in decreased viscosity (Figure 4.14). Several studies have summarized guidelines to minimize polymer degradation, giving limits of oxygen and iron content.

Chemical degradation by free radical generated by the Red/Ox reaction between oxygen and iron or hydrogen sulfide.

Figure 4.14

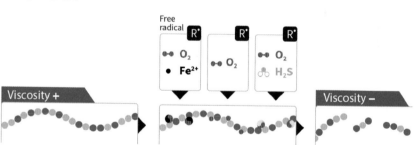

4.5.1.1. Oxygen

Studies have shown that an oxygen concentration below 5 ppb allows good stability of a standard copolymer of acrylamide and acrylic acid up to 120 °C for more than 200 days [15]. Additionally, when the oxygen concentration is 200 ppb or less, minor polymer degradation can be expected when the temperature is low (below 50 °C) and even if significant amounts of Fe^{2+} are present [47]. Another study summarized findings of degradation tests performed at low temperatures (below 50 °C): whatever the level of Fe^{2+}, less than 10% degradation can be expected if the dissolved oxygen content remains below approximately 40 ppb [48]. If the oxygen content in the injection water can't be maintained below acceptable levels, adding an oxygen scavenger such as dithionite, sulfite, or bisulfite can help decrease the free oxygen content. Internal studies have shown that sulfite should be used in fresh water, whereas bisulfite was more efficient in brines.

4.5.1.2. Iron

The presence of Fe^{2+} in the injection water (without oxygen) alone is not detrimental to polymer stability. The impact of Fe^{3+} was also studied to evaluate the formation of gels via crosslinking. However, the vast majority of the tests were carried out in a controlled environment with an excess of $FeCl_3$, favoring the presence of Fe^{3+}. In real field conditions, Fe^{3+} exists as a species predominantly at pH below 2. Above this value, iron (hydr)oxides are dominant in solution and don't contribute to polymer crosslinking.

Questions can arise when the transition occurs from Fe^{2+} to Fe^{3+}. In the laboratory, orange precipitates appear, which are probably a mixture of gels and flocs formed by the precipitation of newly formed iron oxides. In the field, depending on the type and concentrations of contaminants, it is unlikely that crosslinking will occur. However, iron oxides, if present in large quantities, will be flocculated by the polymer injected in the formation, possibly damaging the near wellbore area.

4.5.1.3. Protection from Chemical Degradation

Many strategies can be applied, depending on the case. The simplest scenario is when no oxygen is present, regardless of the presence of other contaminants and their concentrations. In that case, little polymer degradation should be expected. If dissolved oxygen exists in the water, but without any source of Fe^{2+} or H_2S, little to no degradation will occur.

When oxygen and iron are present in the water, two cases can be differentiated for low-to-medium temperature reservoirs:

• If iron is present, but the dissolved oxygen level can be kept below 40 ppb, little degradation should be expected (less than 10%).

• If oxygen levels are above 40 ppb and iron is present, then significant degradation can be expected.

For very high-temperature reservoirs and/or injection water, it is advisable to keep the oxygen level below 5 ppb to prevent any type of redox reaction and free radical generation. Moreover, very low levels of dissolved oxygen are often required to minimize corrosion issues.

If the level of oxygen cannot be maintained below a reasonable level, and if iron or H_2S is present, other strategies can be applied, in order of simplicity and preference:

1) Design specific equipment, and apply strategies to minimize oxygen leaks/ingress and blanket the dissolution facilities, e.g. with nitrogen.

2) Remove the oxygen by adding scavengers (dithionite, sulfite, bisulfite), taking care that no further oxygen leaks occur that would generate a redox reaction with the scavenger and further degrade the polymer.

3) Add free radical scavengers to the polymer itself or in the polymer solution. Many researchers have found that the addition of certain chemicals can stabilize the polymer solution (alcohols, thiourea, sodium borohydride) [49–51]. Such a formulation was used by Petroleum Development Oman (PDO) for the Marmul polymer flood, where a cocktail of isopropanol and thiourea was mixed with the polymer solution to prevent degradation.

4) Remove iron (II) or H_2S. These solutions are by far the most complex and expensive. Removing iron can be achieved by precipitation (aerating the water or changing the pH) and/ or by specific filtration technologies (green sands).

5) Keep the iron under the ferrous state, Fe (II), with reducers (sodium dithionite).

An additional possibility, which can be combined with one of the aforementioned strategies, is to slightly change the polymer chemistry by incorporating ATBS or add specific protective additives [52, 53]; this monomer has shown more resistance to chemical degradation than conventional copolymers of acrylamide and acrylic acid.

Finally, it should be remembered that such degradation can occur during improper polymer sampling or analysis, providing biased viscosity values. Care should be taken to ensure that the proper viscosity value is measured and to adapt the sampling system accordingly. Examples will be provided in Section 5.6.

4.5.2. Mechanical Degradation

Mechanical degradation of the polymer occurs when the molecule is subjected to excessive shear rate or singular pressure drops in a pipe, a choke, an orifice, or a pump, resulting in chain scission and eventually in viscosity loss (Figure 4.15). The higher the molecular weight and chain length, the higher the sensitivity to mechanical degradation. In a typical polymer sample with a certain PDI, the portion with longer chains will be affected first by any excessive shear, resulting in a narrower

Figure 4.15 Mechanical degradation.

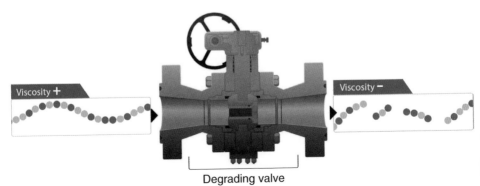

Viscosity +

Viscosity −

Degrading valve

molecular weight distribution. This can be assessed using several tools, including screen factor (SF) and core flooding. Both tests will be described in Section 4.6.7.

Excessive shear may be encountered:

- In the dissolution and injection equipment, if it is not properly designed.

- If fluid velocity (and therefore shear) is too high. For instance, in a 2-in. pipe, it is recommended to keep fluid velocity below $7\,m\,s^{-1}$ to minimize degradation.

- In the completion, if the near wellbore is damaged or if not enough perforations exist. It is recommended to have a minimum of 12 perforations per foot in cased verticals, for instance, and also to clean the wellbore before starting polymer injection.

The choice of equipment is paramount to ensure that little degradation will occur in the system. Following are some examples:

- Centrifugal pumps should not be used for the polymer solution. Triplex or progressive cavity pumps are preferred.

- Manifolds should not be used to control the flow rate. Chokes should be fully open, avoided, or replaced with non-shearing devices.

- Turbine flow meters should be avoided.

Overall, close cooperation between companies is required to optimize the design and ensure that the polymer solution keeps its viscosity throughout the surface equipment, down to the reservoir.

4.5.3. Thermal Degradation

Thermal degradation is often related to an increase in the degree of hydrolysis with respect to temperature and time, and results in molecule precipitation, especially if divalent cations are present in the brine.

The main objective is to select a polymer with an optimized chemistry that is able to sustain viscosity during its propagation in the reservoir. The choice of the right monomer and structure will help minimize possible adverse effects of temperature on the polymer backbone. Usually, long-term stability tests are performed to ensure that minimal viscosity is lost over time. Polymer solutions are prepared in glove boxes and stored in stainless steel ampoules, which are placed in an oven at constant reservoir temperature. Viscosity is periodically checked in total anaerobic conditions.

The main design aspects that can be considered are the following:

- *Polymer chemistry.* The choice of the monomers, structure, and hydrolysis degree is the first step in designing a successful polymer flood. The incorporation of monomers such as ATBS and NVP will prevent hydrolysis and limit viscosity loss over time.

- *Salinity.* The salinity of the injection water is an important consideration. If no divalent cations are present, conventional copolymers of acrylamide and acrylic acid are known to be stable up to 120 °C.

- *Residence time.* The conventional wisdom is to consider the residence time between the injection well and the production well when choosing the appropriate chemistry and designing long-term stability tests. However, this consideration depends greatly on the size of the slug injected and basic reservoir engineering aspects. If only 30% of the reservoir pore volume is filled with polymer, then once water injection is resumed, there is a low probability that the slug will remain intact until the production well. Water will finger through the polymer slug, dilute it, and decrease the efficiency of the entire process; so, a taper of injected polymer concentration and viscosity is commonly practiced and is recommended to minimize injection-side bank deterioration. In theory, the polymer should remain stable only during the duration of injection. This consideration is important because it can help in choosing a chemical composition without overdesigning it.

As with chemistry, examples and guidelines were provided in Section 4.3.2 to help select polymer candidates for further evaluation. In conclusion, we can say that:

- The incorporation of ATBS is required when exceeding 75 °C in mild, saline brines. Depending on the salinity, polymers with ATBS can be stable up to 95 °C.

- NVP can be considered above 100 °C in very salty brines. Polymers with a high level of ATBS are also an alternative for such high temperatures (Figure 4.16).

Long-term stability of different polymer chemistries vs. temperature in synthetic sea water.

Figure 4.16

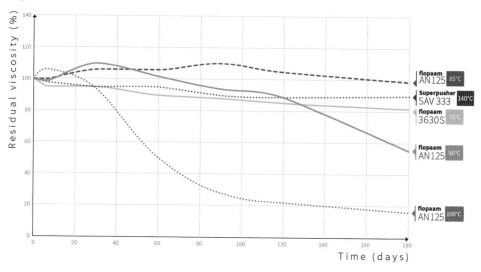

4.5.4. Improving Polymer Stability

As discussed in the previous sections, there are several ways to improve polymer stability.

The first is linked to the polymer itself: choosing the right chemistry, molecular weight, and structure will have an impact on chemical, thermal, and mechanical stabilities. For instance:

- The incorporation of ATBS is known to improve mechanical and thermal stability. Polymers containing ATBS show less viscosity loss when subjected to shear, compared to standard copolymers of acrylamide and acrylic acid.

- A lower-molecular-weight polymer will be less sensitive to mechanical degradation. However, a higher concentration will be required to reach a given target viscosity.

A second strategy could consist of changing the salinity of the injection water or gaining access to a source of fresher water. Lower salinity means more economic chemistry and a lower concentration. In addition, if the reservoir is a good candidate, low salinity or optimized ionic composition has been shown to potentially have a positive impact on oil recovery [54].

A final strategy consists of protecting the polymer from chemical or thermal degradation by adding sacrificial agents or other additives such as thiourea, alcohols, or other previously mentioned methods.

4.6. Laboratory Evaluations

The typical laboratory tests useful for selecting the best polymer candidate will be described in this section. We will try to stick to the most important ones, balancing laboratory findings and interpretations with field applications and results.

Different tests are needed to qualify a powder or an emulsion. The inversion process is critical to the implementation of the emulsion and can be checked with specific devices (flow loop, rheometers measuring torque versus time). The tests described in the following sections apply to both product forms.

4.6.1. Solubility and Filterability

4.6.1.1. Solubility

The ability of the polymer to dissolve, hydrate, and develop viscosity is obviously very important in EOR applications. The equipment and configurations used in a laboratory and in the field are very different; standard procedures are required to make sure the polymer will be readily injectable in the reservoir.

The dissolution of polyacrylamide (both in powder and emulsion) is usually performed with a mechanical stirrer set at 500 rpm (Figure 4.17). It is often recommended to prepare a mother solution @ 0.5% or 1% active concentration to enhance the dissolution efficiency. As recommended by API RP 63 [55], when dissolving powder, the stirring vortex should be maintained and polymer added such that each granule is uniformly wetted. Ideally, the powder should be added quickly enough to introduce the material in less than three minutes, but not so fast that fish eyes are encouraged. For the case of both emulsion and powder blending, the solution is left stirring for 30 minutes at 500 rpm, and then the velocity is decreased to 300 rpm for an additional two hours. The final solution should be free of lumps and undissolved particles before further use. Once this has been completed, a diluted solution can be prepared to obtain the targeted concentration and viscosity.

4.6.1.2. Filterability

The filter ratio (FR) test is commonly used in the industry to check the full dissolution of polymer solutions. It consists of monitoring the filtration rate of a polymer solution when a constant pressure is applied (Figure 4.18). Several procedures exist in the industry, and every company uses a different one, rendering the interpretation of the values obtained sometimes

Figure 4.17 Emulsion inversion process: injection into a vortex.

----

quite difficult. Some of the procedures encountered in the literature are summarized in Table 4.2.

Regarding the type of filter, cellulosic or polycarbonate membrane filters can be used. Internal studies have shown important differences in the final results when comparing both types,

Filter ratio test.

Figure 4.18

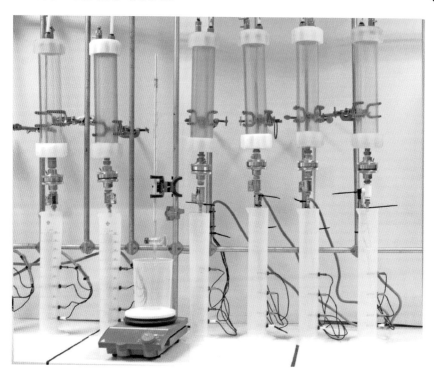

depending on the pore structure and manufacturing processes, causing repeatability and interpretation issues.

It is often necessary to repeat this test two or three times to ensure that the result is representative, especially when the value is close to the target (Figure 4.19). The target FR value should be adapted depending on the reservoir and, especially, its permeability. For screening purposes, a value below 1.5 with a 1.2 µm polycarbonate filter can be set as a target.

4.6.2. Viscosity

The very first requirement when designing a polymer flood is the definition of a target viscosity, or, more precisely, a target resistance factor. The viscosity of several polymers is plotted

Table 4.2 **Examples of filter ratio procedures (from [14]).**

Source(s)	Filter type	Filter diameter	Pore size	Pressure applied	Filter ratio formula	Filter ratio target
[55, 56]	Polycarbonate or cellulosic	47 mm	1.2, 3, and 5 μm	20 psi (1.4 bar)	$FR = \dfrac{t_{500\,ml} - t_{400\,ml}}{t_{200\,ml} - t_{100\,ml}}$	<1.5
[57, 58, 59, 96]	Mainly cellulosic	47 or 90 mm	1.2 μm	1 bar	$FR = \dfrac{t_{200\,ml} - t_{180\,ml}}{t_{80\,ml} - t_{60\,ml}}$	<1.2
[48, 60]	Polycarbonate and cellulosic	47 mm	1.2 and 3 μm	1 bar	$FR = \dfrac{t_{300\,ml} - t_{200\,ml}}{t_{200\,ml} - t_{100\,ml}}$	<1.2

Figure 4.19 **Repeatability of filter ratio tests with different nitrocellulose filters for the same polymer solution.**

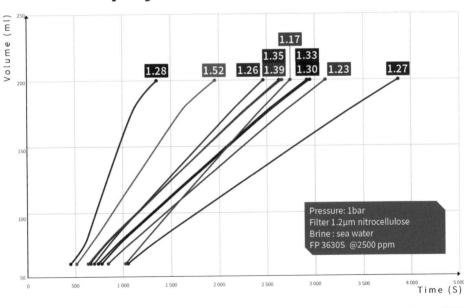

versus concentration in a synthetic field brine at reservoir temperature. By convention, a shear rate of $7.34\,s^{-1}$ is used to measure the values and create the polymer loading curves; it corresponds to the shear rate generated with a Brookfield viscometer equipped with a UL spindle and rotation set at 6 rpm (Figure 4.20). In the field, it is roughly equivalent to the shear obtained for a fluid traveling $1\,ft\,d^{-1}$ in a 25% porosity, 1D

Brookfield viscometer and UL module. Figure 4.20

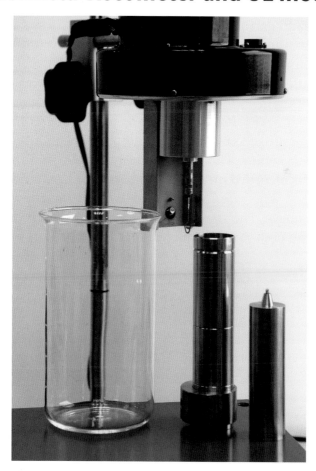

reservoir. Other shear rates can be used depending on the reservoir permeability, wellbore design, and expected injection rates.

This viscosity value will serve as a basis for core flooding, where resistance factor values will be determined at several velocities representative of those observed in the field.

4.6.3. Shear Resistance

In cases where high shear is expected because it is difficult or too expensive to change the design of the surface installations (remove chokes, change the pumps) or recomplete the injection well to add perforations, it is important to assess potential viscosity losses linked to mechanical degradation. Extrapolation from the laboratory to the field is always a challenge, but it is possible to obtain reliable data comparing the shear resistance of different polymers using a capillary shear-degradation test, as described in API RP63 [55]. The basics of the equipment include a pressurized cell connected to a pipe with an internal diameter of 1.75 mm. The pressure drop is fixed, and the fluid throughput is recorded on a scale connected to a computer – to monitor mass. The rate is calculated by plotting the weight-versus-time curve. The viscosity of the collected sample is then measured and compared to the original value before shearing:

$$Deg(\%) = \frac{\mu_{deg} - \mu_o}{\mu_o - \mu_w}$$

where $Deg(\%)$ = percentage of viscosity loss, μ_{deg} = viscosity measured after being sheared, μ_o is the initial viscosity, and μ_w is the viscosity of the water at 20 °C, all in centipoise (cP).

The equivalent shear rate, $\dot{\gamma}$, is calculated using the following equation:

$$\dot{\gamma} = \frac{4q}{\pi r_3} = \frac{4v}{r}$$

where v = superficial velocity (m s^{-1}), r = radius of the capillary (m), and q = flow rate (m^3 s^{-1}).

Although it is relatively easy to calculate the shear through a choke, determining the exact value in the well completion is much more difficult, especially when the flow regime in the near wellbore area is not well understood (radial versus linear flow, microfractures, etc.). Strategies exist to ensure that minimal degradation occurs, as explained in Section 4.5.2: changing the chemistry, changing the molecular weight, adapting the surface installations, fracturing the well, reperforating when applicable, etc.

4.6.4. Screen Factor

The screen factor (SF) test is a common tool to assess the viscoelastic nature of polymer solutions and, more specifically, the impact of high-molecular-weight tails in the polymer molecular weight distribution [61]. The procedure is described in API RP63: it consists of measuring the time for a polymer solution to go through a series of five identically sized screens and comparing it to the solvent baseline flow. Combined with a simple viscosity measurement, this is a good tool to evaluate the impact that long polymer chains will have on resistance to flow.

Internal experiments were carried out, comparing the viscosity and screen factor of polymer solutions before and after passing a simple coil tubing (pressure drop of 1 bar): the viscosity measured before and after were 17.7 and 17.2 cP, respectively (25 °C, 7.34 s^{-1}); therefore, little degradation occurred. For screen factors, the values were 68.2 and 24.4 before and after the coils, respectively. Although very little viscosity loss was observed, a significant change in screen factor indicated that larger molecules were impacted. Therefore, the larger molecules that were sheared appeared to make little contribution to the measured bulk viscosity, but played a more important role for flow through porous media. Practically speaking, this means that larger molecules, being the most sensitive, are degraded relatively quickly; they do not have a large impact on viscosity but prevent possible propagation issues in the field. This aspect should also be investigated when studying the viscoelastic properties of polymers and any possible impact on oil recovery.

4.6.5. Long-Term Stability

Ensuring that viscosity will remain stable during the time necessary to push an oil bank is critical for any chemical EOR project. This is normally achieved by storing carefully prepared polymer solutions in stainless steel ampoules, in an oven at reservoir temperature. The viscosity is checked periodically with a Brookfield viscometer placed in a fully anaerobic environment such as a glove box with nitrogen atmosphere (Figure 4.21). Also, the polymer solutions must be prepared in a confined environment to mimic reservoir conditions and avoid any contaminant ingress that could lead to polymer degradation.

The main question is how long the ampoules should be stored. The answer is usually based on the theoretical residence time of

---

Figure 4.21 **Glove box.**

the polymer in the reservoir, i.e. the time between the injection well and production well. However, this assumes that the polymer slug will remain intact throughout the transit in the reservoir, even when water injection is resumed. But in the vast majority of cases, once polymer injection is stopped and the waterflood is resumed, the water fingers through the polymer slug and dilutes it. At this point, the efficiency of the process is compromised and the polymer slug loses its efficiency. So, as a rule of thumb, stability studies should be performed on a timeframe equal to the residence time multiplied by the percentage of pore volume injected.

Within the glove box, it is also possible to add precise dosages of contaminants to assess their impact on polymer viscosity, mimicking what can happen in the field when the water used to dissolve the polymer contains such species.

4.6.6. Compatibility Tests

When different additives are used in the field (scale inhibitor, corrosion inhibitor, biocide, etc.), it is necessary to ensure that all are compatible with the polymer under consideration for the flood. This is also required when coinjecting a surfactant cocktail along with the polymer in surfactant-polymer (SP) processes. The test consists of preparing the polymer solution at the target concentration and mixing it with the chemicals at the dosage used in the field. The first simple rule to minimize viscosity loss is not to mix a positively charged, cationic product with a negatively charged, anionic synthetic polyacrylamide. A list of compatible biocides follows:

- Acrolein

- Glutaraldehyde

- 2-bromo-2-nitropropane-1,3-diol

- 4,4 dimethyloxazolidine

- 1,2 benzisothiazole-3(2H)-one

- 5-chloro-2-methyl-4-isothiazolin-3-one/2-methyl-4-isothia-zolin-3-one

The biocides listed next are not compatible with anionic HPAM:

- Tetrakis hydroxy methyl phosphonium sulfate (THPS). For instance, a 500 ppm dosage can cause a viscosity loss of 20–30% in two hours.

- Quaternary ammonium-based biocides. This biocide is cationic and precipitates when mixed with the polymer solution.

- Oxidizing biocides such as sodium hypochlorite and chloroisocyanurates.

Scale inhibitors from the phosphonate, phosphinate, and acrylate families are usually compatible with the anionic HPAM.

Corrosion inhibitors are often complex formulations with different molecules. Recent tests have shown that imidazole and methenamine molecules are compatible with anionic HPAM. Performing tests is usually recommended to ensure that no polymer viscosity loss is observed.

4.6.7. Core Flooding

Polymer injection into a porous medium is a good tool to ensure that little formation damage occurs during and after the propagation of the macromolecules. However, it is necessary to define precisely the goals of such experiments and the relevance of the parameters obtained for upscaling or even for pilot implementation.

A good starting point is to define what core floods will not tell:

- Core flooding will never give tangible values for injectivity in the field. None of the near wellbore complexity is taken into account during core flooding: well completion, flow rates, formation damage, microfractures, etc. An example is the shear-thickening effect, which is always studied in the laboratory but will not occur in the field if microfractures are present – which is very often the case in vertical injection wells, where injection rates can be higher and thermal cracking is possible due to cooler injection water. Exceptions to this rule do exist.

- Core flooding will never give true polymer retention values; it only gives a possible range and is more useful for comparative evaluation of each polymer. Each reservoir has a degree of heterogeneity that is not represented during this test.

- Oil recovery experiments should be interpreted with care, and the recovery factor values obtained used with caution for upscaling. A one-dimensional flow experiment in a small piece of core is obviously likely to recover significant volumes of oil and therefore better represents a maximum displacement efficiency if unlimited pore volumes of polymer were available for displacement.

More realistic objectives could be the following:

- Compare several polymers and their in situ flow behavior: primarily, propagation and resistance factor.

- Compare the relative retentions of several polymers and, possibly, the residual resistance factors (permeability reduction linked to the damage caused by the polymer).

- Check the oil recovery with several polymers. It is best if they are injected one after the other in the same core, to check the extra benefits of different chemistries or molecular weights.

- Gather data to be used wisely in 3D simulators.

4.6.7.1. Generalities

Once one or two polymer candidates have been selected, along with a target viscosity, a preliminary injectivity screening can be performed with analogue cores before moving to reservoir cores. The concept is quite simple: a polymer solution is injected with a pump into a piece of rock at several velocities, and the pressure drop is recorded as the solution moves through the core. The pressure drop recorded is then analyzed and compared to those observed with water at comparable flow rates. Darcy's law can be used to calculate two key flow behavior parameters, such as the resistance factor and the residual resistance factor (Figure 4.22). Recall:

$$Q = \frac{k\,\Delta P\,A}{\mu\,L}\ \text{Darcy's law}$$

where Q = flow rate $(\text{m}\,\text{s}^{-1})$, k = permeability (m^2), ΔP = pressure drop $(\text{kg}\,(\text{m}\,\text{s}^{-2})^{-1})$, A = area (m^2), μ = dynamic viscosity $(\text{kg}\,(\text{m}\,\text{s}^{-1})^{-1})$, and L = length (m). Or, in lab units: Q = flow rate $(\text{cm}^3\ \text{s}^{-1})$, k = permeability (Darcy), ΔP = pressure drop $(\text{atm}\,\text{cm}^{-1})$, A = area (cm^2), μ = dynamic viscosity (cP), and L = length (cm).

Figure 4.22 **Rheology of a polyacrylamide solution in a porous medium.**

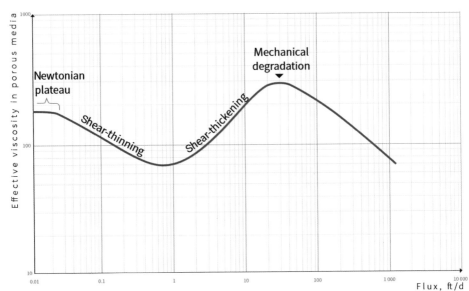

If a core is monophasically saturated with brine and both polymer and brine injections are evaluated at comparable flow rates in the same core flood, then the remaining elements are viscosity and pressure drop. Thus, the concept of resistance factor (Fr), or mobility reduction (R_m) allows us to understand the apparent relative viscosity of the polymer solution during its flow through the porous medium, as defined by:

$$R_m = \frac{\Delta P_{polymer\ solution}}{\Delta P_{brine}} \alpha \frac{\mu_{polymer\ solution}}{\mu_{brine}}$$

where $\Delta P_{polymer\ solution}$ and ΔP_{brine} are the pressure drops during polymer injection and brine, respectively, and μ is the viscosity of the fluids considered.

The permeability reduction or residual resistance factor (R_k) is the reduction of permeability in the presence of adsorbed/lost polymer in the medium. It is defined as:

$$R_k = \alpha \frac{k_{brine}}{k_{post\ polymer}}$$

These relationships can also be determined at residual oil saturation conditions and would therefore correspond to the end-point, resistance factor, or mobility reduction in a two-phase system. However, we must be careful that no additional changes in oil saturation occur during measurement of pressure drops during the equivalent brine and polymer flow rates; otherwise, inaccurate or skewed trends in flow behavior may be observed.

Knowing the aforementioned concepts, we can review the most important points to designing and understanding core floods.

4.6.7.1.1. How Many Core Floods Are Really Necessary?

To answer this question, it is necessary to define the parameters needed for upscaling, as well as the data required to qualify a polymer. In order to minimize the number of experiments and obtain relevant parameters, we can narrow the number to two core flood tests for each representative rock type:

- One reservoir core test at residual oil saturation with dead oil (or live oil) to determine polymer retention at the target concentration (or with another dosage, depending on the method used to calculate the retention value).

- One reservoir core test at residual oil saturation to determine resistance factors versus velocity for the target polymer concentration, starting with the velocity expected deep in the reservoir. The objective is to obtain a plot of R_m versus velocity (or shear rate). At the end of this process, it is also possible to determine a residual resistance factor if the reservoir has a low permeability.

If reservoir cores are unavailable or limited, appropriate outcrop cores with representative permeability, porosity, and mineralogy may be used for screening.

The choice of the core is important and should be dictated by the facies distribution in the reservoir. Given a heterogeneous reservoir with two main layers, one having an average permeability of 1000 mD and the other of 300 mD, the choice of the polymer molecular weight and the core depends on the proportions. If 80% of the reservoir has an average permeability of 1000 mD, while 20% of the reservoir exhibits a permeability of 300 mD, the design should be based on the higher permeability. Obviously, this also depends on oil saturation distribution, reservoir configuration, crossflow, and/or the presence of barriers between the layers. If it is technically possible and economically viable to isolate both zones and inject different polymers and/or viscosities, then this would be the ideal strategy to exploit.

4.6.7.1.2. What Velocities Should Be Considered?

It is paramount to determine a velocity map in the reservoir and pinpoint an average velocity deep inside the formation where the polymer will mobilize the oil bank. Quite often, the velocity is very low (producing shear rates $<10\,s^{-1}$) and the flow is found to be laminar, and the viscosity of the solution can be easily determined by looking at the low-shear Newtonian plateau on a plot of viscosity versus shear rate. For a core flood test, several velocities can be considered to gather data for 3D simulation and upscaling. However, great care should be taken when considering high velocities, since many observations made in the laboratory will not be applicable in the field – especially when dealing with the near wellbore area (shear-thickening or pseudo-dilatant behavior, for example). This aspect will be discussed in Section 5.3.

4.6.7.1.3. Is an Oil Recovery Test Relevant?

For straight polymer flooding, no. Injecting a viscous solution after a waterflood in a small piece of rock with a unidirectional flow will always recover more oil, by definition. This type of experiment is convenient to prove a concept; but, very often, the results of oil recovery tests are history-matched with simulators and used for upscaling and, finally, to build a business case. In a core, it is relatively easy to obtain an additional 30% oil recovery, but this is most likely not possible in the field. Conversely, building a core flood-based test on a field design is also questionable: injecting 30% of core pore volume with polymer, flushing with water, and calculating recovery misses several points:

- A small core is not similar to the entire reservoir. Heterogeneities are missing, as are crossflow, 3D dimensional flow, etc. This has a huge impact on oil recovery.

- Wettability, saturations, and fluid properties are too different between the laboratory and the field.

- In a small model, the displacement characteristics are much more favorable than in the field.

If oil recovery tests are considered, it is recommended to do the following:

- *Consider a core with reasonable dimensions.* It should be a minimum of 1-ft long (30 cm) and 1.5″ (3.81 cm) in diameter (for reservoirs with limited cores, a multi-plug, composite core might be required).

- *Adapt the injection history to the field.* Inject water until breakthrough or matching the field water-cut, and then switch to polymer until no oil production is observed. Finally, switch back to water.

- Consider different polymer viscosities to assess the impact on oil recovery.

- Use the data gathered with care for upscaling.

Oil recovery tests are important when designing SP or alkali-surfactant-polymer (ASP), checking the efficiency of the formulation and the impact on oil recovery and final saturations.

4.6.7.1.4. How Do We Consider the Residual Resistance Factor Value?

The residual resistance factor, R_k, is calculated by dividing the pressure drop during water/brine chase after polymer by the pressure drop of the water/brine baseline. This provides an indication of the damage caused to the core by the injection of the polymer solution. For instance, an R_k value of 2 indicates that the initial permeability to water/brine was divided by 2 or is ½ the original permeability. However, there are several problems:

- When switching from polymer to water chase, a new, unstable displacement front is created with two miscible fluids. Therefore, it will take a very long time and many pore volumes before the water can really flush out all the polymer. A way to minimize this issue is to progressively decrease the slug viscosity before switching back to water. But even in this case, several hundreds of pore volumes may be necessary to flush out all the polymer (it can be confirmed by analyzing the effluents and checking for the presence of the polymer).

- Given a polymer solution that enters all reservoir zones, the possible damage will be more important in low-permeability zones than in high-permeability zones. However, as mentioned, small cores exhibit limited heterogeneity and do not represent the entire reservoir. In the field, when water injection is resumed, water will again preferentially finger through high-permeability zones and not evenly displace the polymer slug toward the producer. In that case, calculating a single residual resistance factor value makes little sense. In high-permeability zones preferentially flushed by water, the value will likely be 1.

In the literature, a high residual resistance factor is generally considered to be good for diverting chase water to previously unswept zones, especially when very large permeability contrasts exist where channeling dominates the flow profile. However, for injectors with less noticeable permeability contrasts, this may also mean reduced injectivity, high retention, and capacity for only lower-viscosity injected polymer solution to displace the oil, which is the main parameter behind the success (or failure) of viscous water injection.

For simulation purposes, and given the previous considerations, it is recommended to use an R_k value of 1 in simulations, with some exceptions:

- In reservoirs with permeability below 200–300 mD, it is necessary to quantify the possible damage.

- When associative polymers are considered, resistance factors and residual resistance factors are important parameters, given the characteristics of such products.

4.6.7.1.5. What Is the Best Way to Determine Polymer Retention and Minimize Retention?

Polymers are macromolecules that can interact with the rock when they flow through a reservoir. Usually, a small percentage of the injected polymer fluid is lost due to retention. This encompasses three mechanisms:

- *Adsorption.* The polymer "sticks" to the rock via van der Waals forces or ionic or hydrogen bonds. Given the possibility for the molecule to attach to the rock at many points, adsorption is usually considered irreversible.

- *Mechanical entrapment.* The very large molecules can become physically trapped at the entrance of pore throats.

- *Hydrodynamic retention.* Molecules can be temporarily trapped in regions where the flow is stagnant.

The flow rate can have an impact of the last two mechanisms, but, generally speaking, polymer retention can be considered nearly irreversible.

In the laboratory, polymer retention is determined either using a static or a dynamic method. The static method consists of mixing a known mass of polymer with a known mass of crushed reservoir rock or sand and, after settling and gentle agitation, determining the polymer concentration in the supernatant by mass balance. This method has several obvious limits. First, only physical adsorption is determined by this method; the in situ behavior is not taken into consideration. Second, the area available for adsorption is much higher for free sand grains than for consolidated reservoir rock, yielding high adsorption values. Third, polymer can stick to the vessel used for the test, skewing the concentration values of the supernatant determined by mass balance.

The preferred technique to quantify polymer retention is the dynamic method (Figure 4.23). In this case, two banks of polymer solution are injected, separated by a brine slug. A tracer can also be added to determine inaccessible pore volume

Figure 4.23 Determining retention and inaccessible pore volume with the dynamic method (two fronts).

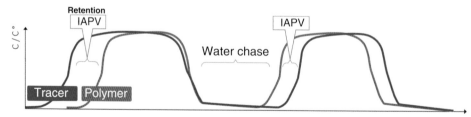

by comparing the breakthrough times between the tracer (potassium iodide, for example, detected with a UV spectrophotometer) and the polymer. Polymer retention can be determined by plotting the concentration or viscosity profiles of the effluents versus pore volume injected [62].

In the first case, a fraction collector is used to collect and analyze the effluent samples, such that a produced polymer concentration profile can be obtained. Several methods can be used for to quantify the polymer residuals:

- Starch iodide titration method.

- Bleach titration method.

- Total organic carbon analyzer coupled with a nitrogen-detection module, to differentiate between other components (e.g. when oil is present).

- Colloidal titration method (limited when salinity is important).

The starch iodide and bleach titration methods require specific sample quality to avoid interference with other contaminants when evaluating the absorbance at specified wavelengths to determine the polymer concentration.

It is also possible to determine retention by plotting the viscosity profile if an inline capillary viscometer is placed at the outlet of the core. In that case, retention is determined by calculating the area between the viscosity curves of the first and second front. Both methods (dosage and viscosity) can be combined to increase the accuracy of the measurement. Practically speaking, retention values below $50\,\mu g\,g^{-1}$ are acceptable. The value can be impacted by various parameters, including the following:

- Polymer chemistry

- Molecular weight

- Salinity

- Concentration injected

- Lithology, presence of clays

When a high retention value is measured, it is possible to play on the aforementioned parameters to favor polymer propagation:

- *Change the polymer chemistry.* For instance, incorporating ATBS is known to decrease retention.

- Decrease the molecular weight to minimize mechanical and hydrodynamic entrapment.

- Change the salinity to minimize ionic bonding.

- Inject sacrificial agents to presaturate the adsorption sites.

It was proposed to inject sacrificial agents before the chemical cocktail to saturate the adsorption sites and minimize retention of the polymer and/or surfactants injected in chemical EOR processes [63, 64]. This strategy can be efficient but also costly. In theory, to be really effective, the sacrificial agent should be injected over the same pore volume (specifically, the reservoir mass, since it is a mass balance) as the subsequent chemical cocktail. Or it should be designed so that a very high concentration is injected first, which, by dilution, will saturate the rock while propagating; but this case is more difficult to design and poses other problems. The simplest way, for polymer injection, would be to slightly overdose the polymer to compensate for losses in the reservoir (e.g. injecting 1050 ppm instead of 1000 ppm). This strategy helps avoid handling several additional chemicals.

4.6.7.1.6 What Is the Best Way to Check Polymer Viscosity and Effluent Quality?

Measuring the viscosity of the effluents is critical, to check that no polymer degradation occurred through the core, especially when setting velocity limits to quantify mechanical degradation. It is recommended to avoid measuring the viscosity of the collected effluents, since any oxygen ingress can induce chemical degradation and bias the measurement, especially when reservoir fluids are used. The best way to determine polymer viscosity and quality is to install an inline capillary viscometer at the outlet of the core, to calculate the relative viscosity at a given velocity and shear rate, and correlate this value with a viscosity versus shear rate profile obtained with conventional rheometers.

4.6.7.1.7 What About Polymer Viscoelasticity?

Several authors have noted that the use of viscoelastic polymers helped recover oil from cores beyond what was expected from a simple increase of aqueous-phase viscosity. Many studies have been carried out so far, with different conclusions, making a generalization quite difficult [65–73]. Among the proposed mechanisms, elastic turbulence causing breakup and mobilization of trapped oil ganglia was investigated. Experiments were carried out with high-molecular-weight polymers (>15 million Da), suggesting that their intrinsic properties (long relaxation time) were partly responsible for the recovery of additional oil (ibid). In addition, the studies showed that the results were also rate- and concentration-dependent. The higher the polymer concentration, its molecular weight, and the injection rate, the better the results. The question that arises is whether this effect will be observed in the field, knowing that:

- The shear rate decreases very quickly in the reservoir.

- The polymer concentration is affected by retention, with lower concentration at the front.

- The high-molecular-weight fraction will be the first impacted by mechanical degradation, removing the long, elastic molecules first.

From a practical standpoint, many uncertainties remain about applicability in the field: polymer viscoelasticity should be incorporated in a polymer flood design as the cherry on the cake, with efforts focused on the resistance factor and propagation.

4.6.7.2. Equipment and Tips for Injection

The injection of polymer solutions in a porous medium requires specific care to minimize degradation. Some guidelines are provided next:

- Brines and polymer solutions should be prepared in anaerobic conditions, preferably in a glove box. The vessels or cylinders used to inject the solutions should also be filled in the glove box and sealed.

- All parts that come in contact with the fluids should be composed of Monel or Hastelloy. All pipes conducting the fluids should be in stainless steel or other alloys. Nylon should be avoided, to minimize oxygen diffusion, especially at high temperatures, but also to increase tolerance to pressure spikes during polymer propagation and rate changes.

- An inline oxygen probe is always a good tool to ensure that the system is leak-free. It was observed that, even with new systems, small leaks could greatly impact the solution viscosity, given the residence time in the system, as well as the temperature when dealing with hot reservoirs. It is therefore recommended to systematically add a protective agent such as thiourea in the brine and polymer solutions to prevent unexpected chemical degradation. It is possible to add other probes: pH, conductivity, etc.

- An inline capillary viscometer should be placed at the outlet of the core to measure the viscosity without any degradation. With a core bypass, it is even possible to run viscosity measurements directly in the capillary, even prior to injection into the core. Values can be correlated with a viscosity profile obtained in the laboratory to confirm them before starting the core flood test.

- Core holders with multiple pressure taps are preferred to check propagation over the entire core and differentiate between face plugging and other phenomena when necessary.

In a nutshell, it is necessary to carefully prepare the polymer solution and ensure that the system is free of any leaks when performing a core flood. Pressure sensors should be calibrated routinely, because many of the calculations depend on accurate and consistent pressure measurement. The addition of protective additives is a good way to prevent degradation, especially at high temperatures. Finally, the most important parameters for running a 3D simulation and building a business case are retention and resistance factors (at different velocities). It is strongly advised not to consider a residual resistance factor other than 1, in order for good-quality reservoirs to build a business case and calculate oil recovery; too many uncertainties will render the process very complex.

4.6.8. Quality Control

Routine procedures exist to check the quality of polymer batches. The parameters are listed in a certificate of analysis delivered with the polymer, with reference to the specific procedure used. The main parameters are as follows:

- *Solid content of the powder* (determining the active content)
- *Residual acrylamide* (checking for free residual monomer)
- UL viscosity
- *Insolubles* (checking for possible fish eyes or dissolution issues)

In some cases, the filterability specifications and the yield viscosity can be added to the certificate of analysis. The yield viscosity corresponds to the concentration needed to reach the target viscosity in a specific brine at a given temperature and shear rate. An example of specifications for a standard 30% ionic copolymer of acrylamide and acrylic acid in powder form with very high molecular weight is given in (Table 4.3).

Some of the laboratory procedures commonly used are described in API RP63. Discussions with suppliers can also

Table 4.3

Example of basic specification for quality control of polymer in powder form

Parameter	Unit	Specification
Total solids	%	88–100
Residual acrylamide	ppm	0–999
Insoluble	%	0–1
UL viscosity	cP	6.50–7.50
Solution appearance		Slightly hazy solution

provide important input regarding the proper handling of products and control of overall quality.

In the field, the main parameter to be monitored is polymer solution viscosity, which can be measured via an inline viscometer or by manual and careful sampling to avoid chemical and mechanical degradation.

4.6.9. Heath, Safety, and Environment

Two aspects are important when discussing polyacrylamide-related hazards. The first is directly linked to product handling for both powder and emulsion forms. The second is related to the polymer and its fate in the environment.

4.6.9.1 Product Handling

Both emulsion and powder forms have a limited shelf life (six months and one year, respectively, with proper storage conditions). The only hazard associated with the handling, use, and storage of both emulsion and powder products is a *slip hazard*. These products – emulsions as supplied, and powders when wet – render surfaces extremely slippery. All spills, no matter how small, must be cleaned up promptly.

4.6.9.2 Anionic Polyacrylamide in the Marine Environment

Anionic polyacrylamide comes with all the data required to complete the Harmonized Offshore Chemical Notification Format (HOCNF) for use in offshore applications in the North Sea. The results obtained in aquatic toxicity tests demonstrate a very low order of toxicity (studies to be published). In the sediment reworker test, for example, the number of immobilized amphipods did not exceed 5% relative to control at the highest dose tested ($11\,984\,mg\,kg^{-1}$), which is the limit of solubility due to viscosity constraints. In other tests (fish, crustaceans, and algae), effects were seen only at very high concentrations, and the effect was probably due to the significantly increased viscosity of the test medium induced by such high concentrations of the polymer. A $1\,g\,l^{-1}$ (1000 ppm) solution in seawater – approximately equivalent to the no effect concentration (NEC) – has a viscosity around 10 cps, 10 times greater than that of seawater. Additionally, due to the design of the test, crustaceans are more sensitive to viscosity, since it causes the test organisms to stick together and to the vessel walls and reduces the oxygen partial pressure. This is especially true when the endpoint is immobilization.

Anionic polyacrylamide demonstrates no toxicity in acute or long-term mammalian tests [74–81].

The octanol-water partition coefficient and the resulting Bioconcentration Factor (BCF) are very low, excluding any potential for the polymer to bioconcentrate/bioaccumulate. The polymer does not, however, demonstrate biodegradability in the standardized tests.

The back-produced anionic polyacrylamide will already be significantly degraded due to the conditions in the production process (pumps) and other possible physical impacts in the reservoir (hydrolysis, chromatographic separation). If less than 1 reservoir pore volume is injected, then the injected concentration will never be reproduced because of dilution by the water chase. Also, when discharged to the sea, it will be diluted to concentrations many orders of magnitude below its NEC. In the aquatic medium, it will be exposed to naturally occurring hydroxyl radicals, metal ions, and ultraviolet light, all of which degrade the molecular weight in a continuous manner. Anionic polyacrylamide has a high propensity to adsorb irreversibly to positively charged and neutral particles suspended in the water column. Therefore, when the polymer (or a degraded fraction) encounters such particles in the marine environment, it will adsorb to them, precipitate out of solution, and then slowly sink to the bottom, where, as demonstrated in the sediment reworker test (at concentrations as high as $10\,g\,kg^{-1}$), there will be no negative impact.

4.6.9.3. Biodegradability

High-molecular-weight vinyl polymers are resistant to microbial degradation in standardized biodegradation tests such as the OECD 306 (Organization for Economic Cooperation and Development). This is due to the stability of both the polymer chain and the functional lateral groups (which also result in the non-toxicity of the polymer, the molecule being too large to pass biological membranes). The polymer's size and stability in the aqueous medium, combined with the electrostatic surface activity (the polymer has a negative charge), prevent bacteria from using it as a ready source of nutrition. Extending the test duration (as per the marine biodegradability of insoluble substances [BODIS] test) does not improve the situation, since there are no environmental factors present in the test capable of reducing the size or the surface charge. Deprived of nutrition, the bacteria die off.

However, these carbon-carbon polymers (i.e. the polymer backbone is made up of single-bonded carbon atoms) can be degraded physically and chemically [82–90]. Molecular weight is reduced by shear and heat (both encountered somewhere in the injection process) and by radicals, especially hydroxyl radicals ($\cdot OH$), metal ions such as Cu^{2+} and Fe^{3+}, and UV light present in significant concentrations in the marine environment [91].

The polymer present in the produced water will have been degraded during its passage through the reservoir and will then be exposed to UV light at up to 20 m below the surface and hydroxyl radical fluxes and metal ions at all depths. These will break the polymer chain into smaller and smaller fractions until it reaches a size where it can be degraded biologically (less than 2000–3000 Da).

Like the parent polymer, these degraded fractions have no systemic toxicity, and dilution removes the potential for physical effects. The low reactivity of the polymer – both the polymer chain and the lateral groups – excludes the possibility of recombination to create toxic substances or render them bioavailable.

Many studies have been performed to identify microorganisms able to biodegrade the polyacrylamides by using them either as a nitrogen or a carbon source [82–91]. An extensive review of possible degrading organisms for polyacrylamide, possible mechanisms, and thermodynamics is given in Caiyun et al. [5].

4.6.9.4. Polyacrylamides as a Nitrogen Source

A mechanism has been proposed by Caulfield et al. [92]: an extracellular amidase enzyme hydrolyses the amide groups to form NH_3 and a carboxylic group on the polymer backbone (Figure 4.24). But, in that mechanism, the molecular weight is not changed: biodegradability is not enhanced. The backbone is not used as a nutrient; only a change in chemical composition occurs.

4.6.9.5. Polyacrylamides as a Carbon Source

There is little evidence of the existence of microorganisms that can use the backbone as nutrient [5]. Studies have been performed, but the results are often arguable:

- Bacteria growth was observed in some cases, but HPAM was not the only source of nutrient; glucose was often added.

Figure 4.24 Proposed mechanism for the use of polyacrylamide as a nitrogen source [92].

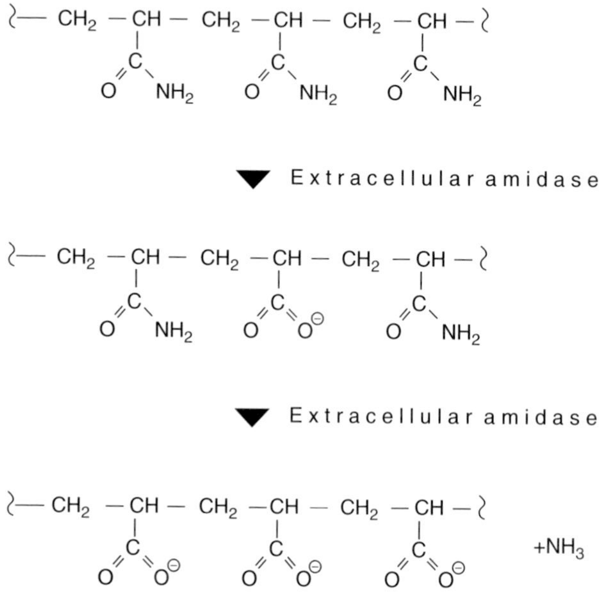

- The analytical means used to quantify PAM consumption are not convincing. In one case, the starch iodide method was used to dose the polymer, but this method detects amide function and is irrelevant to detect the nitrogen components. In another case, IR spectrums of the polymer molecules were compared but no difference was made between the initial nondegraded PAM and the possible degraded molecules.

Another study combined photolytic degradation to reduce the molecular weight and the incorporation of microorganisms to enhance biodegradability [93]. Experiments are still ongoing to clarify the possibility for certain microorganisms to use the polymer backbone as nutrient.

4.6.9.6. About Acrylamide Reformation and Toxicity

Polyacrylamides are long-chain polymers made by polymerization of the acrylamide monomer. A recurring question is whether the polymer can break down (or *unzip*) so as to reform acrylamide monomer, which would be released to the environment.

Acrylamide is very readily biodegradable: i.e. it is a source of nutrition for microorganisms (see [94] and associated references).

The reason polyacrylamide does not break down to reform acrylamide is quite simple: acrylamide contains a double bond, and it is through the breaking of this double bond that polymerization is achieved. This is an exothermic event.

When polymerization is complete, there are no double bonds in the polymer chain, and there is no acrylamide (except trace concentrations that did not react; they can be quantified and are always less than 0.1% of the active polymer).

To reform a double bond requires very significant amounts of energy. A publication by Reber et al. [95] summarizes some interesting findings:

"With regard to the degradation of polyacrylamide to acrylamide monomer, we were able to draw several conclusions. First, is the formation of the first degradation radical which requires the cleavage of carbon–carbon bonds in the polymer chain (72.6–96.6 kcal; approximately 350–550 °C). After the formation of the initial radicals the predominant degradation pathway would be back-biting. This pathway leads to oligomeric degradation products, not acrylamide monomers. Second, the back-biting pathway involves a five- or six-membered ring forming intermediate such that both kinetic and thermodynamic considerations favor this pathway. Third, the pathway that leads to the release of the monomer requires at least 20 kcal which is a relatively high barrier in most potential thermal situations (1 part in 10^{14} at room temperature)."

In a Nutshell

Polymers for EOR are selected based on reservoir temperature, permeability, and the salinity of the injection water. Many variations in chemistry and characteristics are possible to ensure that the molecule will suffer from minimum degradation during its journey from the surface to the reservoir. With the use of careful controls, this nontoxic macromolecule will efficiently viscosify the water and enable the production of additional hydrocarbons.

References

[1] (2004). *Kirk-Othmer Encyclopedia of Chemical Technology*, 5e, vol. 1, A–Ai. Wiley Interscience publication. ISBN: 0-471-48522-5.

[2] Remp, P. and Merrill, E.W. (1991). *Polymer Synthesis*, Chap. 4, 2e. Hülthig & Wepf. ISBN: 3-85739114-6.

[3] Perry, R.H. and Green, D.W. (2007). *Perry's Chemical Engineers' Handbook*, 8e. The McGraw Hill Company, section 17-17. ISBN: 978-0-07-142294-9.

[4] Goempel, K., Tedsen, L., Ruenz, M. et al. (2017). Biomarker monitoring of controlled dietary acrylamide exposure indicates consistent human endogenous background. *Toxicokinetics and Metabolism* https://doi.org/10.1007/s00204-017-1990-1.

[5] Caiyun, L., Hao, H., Jinfeng, L. et al. (2016). Research Progress of polyacrylamide biodegradation. *Oilfield Chemistry* 33 (3).

[6] Yocum, R.H. and Nyquist, E.B. (1973). *Functional Monomers*, Their Preparation, Polymerization, and Application, vol. 1. Ed. Marcel Dekker. 715pp. ISBN: 0-8247-1810-0.

[7] Kulawardana E., Koh H, Kim D. et al. 2012. Rheology and transport of improved eor polymers under harsh reservoir conditions. Paper SPE 154294 presented at the Eighteenth SPE Improved Oil Recovery Symposium, Tulsa, Oklahoma, USA, 14–18 April.

[8] Vermolen E.C.M., Van Haasterecht M.J.T., Masalmeh S.K. et al. 2011. Pushing the envelope for polymer flooding towards high-temperature and high-salinity reservoirs with polyacrylamide based ter-polymers. Paper SPE 141497 presented at SPE Middle East Oil and Gas Show and Conference, Manama, Bahrain, 25–28 September.

[9] Gaillard N., Giovannetti B., Favero C., 2010. Improved oil recovery using thermally and chemically stable compositions based on co and ter-polymers containing acrylamide. Paper SPE 129756 presented at the SPE Improved Oil Recovery Symposium, Tulsa, Oklahoma, USA, 24–28 April.

[10] Fernandez, I.J. 2005. Evaluation of Cationic Water-Soluble Polymers With Improved Thermal Stability. Paper SPE 93003 presented at the SPE International Symposium on Oilfield Chemistry, The Woodlands, Texas, USA, 2–4 February.

[11] Doe, P.H., Moraghi-Araghi, A., Shaw, J.E., and Stahl, G.A. (1987). Development and evaluation of EOR polymers suitable for hostile

environments-Part 1: copolymers of vinylpyrrolidone and acrylamide. *SPE Reservoir Engineering* 2: 461–467.

[12] Favero, C., Gaillard, N., and Giovannetti, B. (2014). Polymers for enhanced hydrocarbon recovery. World patent WO2014166858 filed 7 April 2014 and issued 16 October 2014.

[13] Molyneux, P. (1982). *Water-Soluble Synthetic Polymers: Properties and Behavior*. Chap. III – 3, vol. I. CRC Press. ISBN: 0-8493-6136-2.

[14] Thomas A., Braun O., Dutilleul J. et al. 2017. Design, characterization and implementation of emulsion-based polyacrylamides for chemical enhanced oil recovery Paper EAGE presented at the 19th European Symposium on Improved Oil Recovery, Stavanger, Norway, 24 April. https://doi.org/10.3997/2214-4609.201700286.

[15] Seright, R.S., Campbell, A.R., Mozley, P.S., and Han, P. (2010). Stability of partially hydrolyzed polyacrylamides at elevated temperatures in the absence of divalent cations. *SPE Journal* 341–348.

[16] Audibert, A. and Argillier, J.-F. 1995. Thermal stability of sulfonated polymers. SPE 28953 presented at the SPE International Symposium on Oilfield Chemistry, San Antonio, Texas, USA, 14–17 February.

[17] Gaillard N., Giovannetti B., Favero C. et al. 2014, New water soluble NVP acrylamide terpolymers for use in eor in harsh conditions. SPE 169108 presented at the SPE Improved Oil Recovery Symposium, Tulsa, Oklahoma, USA, 12–16 April.

[18] Hollander, A.F., Ssomasundaran, P., and Gryte, C.C. (1981). Adsorption characteristics of polyacrylamide and sulfonate-containing polyacrylamide copolymers on sodium kaolinite. *Journal of Applied Polymer Science* 26: 2123–2138.

[19] Kamal, M.S., Sultan, A.S., Al-Mubaiyedh, U.A., and Hussein, I.A. (2015). Review on polymer flooding: rheology, adsorption, stability, and field applications of various polymer systems. *Polymer Reviews* 55 (3): 491–530. https://doi.org/10.1080/15583724.2014.982821.

[20] Levitt, D.B. and Pope, G.A. 2008, Selection and screening of polymers for enhanced-oil recovery. SPE113845 presented at the SPE Symposium on Improved Oil Recovery, Tulsa, Oklahoma, USA, 19–23 April.

[21] Moradi-Araghi, A., Cleveland, D.H., Jones, W.W. et al. 1987. Development and evaluation of EOR polymers suitable for hostile environments: II-copolymers of acrylamide and sodium AMPS. Paper SPE 16273 presented at the International Symposium on Oilfield Chemistry, San Antonio, Texas, USA, 4–6 February.

[22] Noïk Ch., Audibert A., and Delaplace Ph. 1994. Injectivity of sulfonated polymers under North Sea oil field conditions. SPE 27769 presented at the SPE/DOE 9th Symposium on Improved Oil Recovery, Tulsa, Oklahoma, USA, 17–20 April.

[23] Parker, W.O. and Lezzi, A. (1993). Hydrolysis of sodium-2-acrylamido-2-methylpropanesulfonate copolymers at elevated temperature in aqueous solution via 13C n.m.r. spectroscopy. *Polymer* 34 (23): 4913–4918.

[24] Ryles, R.G. (1988). Chemical stability limits of water-soluble polymers used in oil recovery. *SPE Reservoir Engineering* 3 (1): 23–34.

[25] Rashidi, M. 2010. Physico-chemistry characterization of sulfonated polyacrylamide polymers for use in polymer flooding. PhD thesis. University of Bergen.

[26] Rashidi, M., Blokhus, A.M., and Skauge, A. (2010). Viscosity and retention of Sulfonated polyacrylamide polymers at high temperature. *Journal of Applied Polymer Science*, published in Wiley InterScience, https://doi.org/10.1002/app.33056.

[27] Rashidi, M., Blokhus, A.M., and Skauge, A. (2009). Viscosity study of salt tolerant polymers. *Journal of Applied Polymer Science*, Published in Wiley InterScience. https://doi.org/10.1002/app.32011.

[28] Zaitoun A., Makakou P., Blin N. et al. 2011. Shear stability of EOR polymers. SPE 141113 presented at the SPE International Symposium on Oilfield Chemistry, The Woodlands, Texas, USA, 11–13 April.

[29] Argabright, P.A., Rhudy, J.S., and Phillips, B.L. (1982). Partially hydrolysed polyacrylamides with superior flooding and injection properties. SPE 11208, presented at the SPE 57th Annual Fall Conference, New Orleans, Louisiana, USA, 26–29 September.

[30] Holzwarth, G., Soni, L., Schulz, D.N., and Bock, J. (1988). Absolute MWDs of polyacrylamides by sedimentation and light scattering. In: *Water Soluble Polymers for Petroleum Recovery* (ed. G.A. Stahl and D.N. Schulz), 215–229. New York: Plenum Publishing Corp.

[31] Gaillard N., Thomas A., Bataille S. et al. 2017. Advanced selection of polymers for eor considering shear and hardness tolerance properties. EAGE31648 presented at the 19th European Symposium on Improved Oil Recovery, Stavanger, Norway, 24–27 April.

[32] Bock, J., Valint, P.L., Pace, S.J., and Gardner, G. 1987. Enhanced oil recoverywith hydrophobically associating polymers containing sulfonate functionality. US Patent No. 4,702,319.

[33] Bock, J., Valint, P.L., Pace, S.J. et al. (1988). Hydrophobically associating polymers. In: *Water-Soluble Polymers for Petroleum Recovery* (ed. G.A. Stahl and D.N. Schulz), 147–160. New York City: Plenum Press.

[34] Buchgraber, M., Clements, T., Castanier, L.M. et al. 2009. The displacement of viscous oil by associative polymer solutions Paper SPE 122400 presented at the SPE Annual Technical Conference and Exhibition, New Orleans, Louisiana, USA, 4–7 October. https://doi.org/10.2118/122400-MS.

[35] Dupuis, G., Rousseau, D., Tabary, R., and Grassl, B. (2011). Flow of hydrophobically modified water-soluble-polymer solutions in porous media: new experimental insights in the diluted regime. *Society of Petroleum Engineers Journal* 16 (1): 43–54. SPE-129884-PA. https://doi.org/10.2118/129884-PA.

[36] Evani, S. 1981. Water-dispersible hydrophobic thickening agent. US Patent 4,432,881, filed 6 February 1981 and issued 21 February 1984.

[37] Evani, S. 1983. Enhanced oil recovery process using a hydrophobic associative composition containing a hydrophilic/hydrophobic polymer. US Patent 4,814,096, filed 1 August 1983 and issued 21 March 1989.

[38] Gaillard, N. and Favero, C. 2007. High molecular weight associative amphoteric polymers and uses thereof. US Patent 7,700,702, filed 13 December 2007 and issued 20 April 2010.

[39] McCormick, C.L. and Johnson, C.B. (1988). Structurally tailored macromolecules for mobility control in enhanced oil recovery. In: *Water-Soluble Polymers for Petroleum Recovery* (ed. G.A. Stahl and D.N. Schulz), 161–180. New York City: Plenum Press.

[40] Taylor, K.C. and Nasr-El-Din, H.A. (1998). Water-soluble hydrophobically associating polymers for improved oil recovery: a literature review. *Journal of Petroleum Science and Engineering* 19 (3–4): 265–280. https://doi.org/10.1016/S0920-4105(97)00048-X.

[41] Taylor, K.C. and Nasr-El-Din, H.A. 2007. Hydrophobically associating polymers for oil field applications. Paper CIPC 2007–016 presented at the Canadian International Petroleum Conference, Calgary, Canada, 12–14 June. https://doi.org/10.2118/2007-016.

[42] Seright, R.S., Fan, T., Wavrik, K. et al. 2011. Rheology of a new sulfonic associative polymer in porous media. Paper SPE 141355 presented at the SPE International Symposium on Oilfield Chemistry, The Woodlands, Texas, USA, 11–13 April.

[43] Winnik, F.M. (1989). Association of hydrophobic polymers in water: fluorescence studies with labeled (hydroxypropyl)celluloses. *Macromolecules* 22 (2): 734–742.

[44] Bokias, G., Hourdet, D., and Iliopoulos, I. (1997). Hydrophobic interactions of poly (N-isopropylacrylamide) with hydrophobically modified poly (sodium acrylate) in aqueous solution. *Macromolecules* 26: 8293.

[45] Hara, M. (1992). *Polyelectrolytes: Science and Technology*. Chap 3. Marcel Dekker. ISBN: 0-8247-8759-5.

[46] Bonnier, J., Rivas, C., Gathier, F. et al. Inline viscosity monitoring of polymer solutions injected in chemical enhanced oil recovery processes Paper SPE 165249 presented at the SPE Enhanced Oil Recovery Conference, Kuala Lumpur, Malaysia, 2–4 July.

[47] Seright, R.S. and Skjevrak, I.. 2014. Effect of dissolved iron and oxygen on stability of HPAM polymers. Paper SPE 169030 presented at the SPE Improved Oil Recovery Symposium, Tulsa, Oklahoma, USA, 12–16 April.

[48] Jouenne, S., Klimenko, A., and Levitt, D. 2016. Polymer flooding: establishing specifications for dissolved oxygen and iron in injection water. Paper SPE179614 presented at the SPE Improved Oil Recovery Conference, Tulsa, Oklahoma, USA, 11–13 April.

[49] Shupe, R.D. (1981). Chemical stability of polyacrylamide polymers. *Journal of Petroleum Technology* 33 (8): 1513–1529. SPE-9299-PA. https://doi.org/10.2118/9299-PA.

[50] Wellington, S.L. (1983). Biopolymer solution viscosity stabilization-polymer degradation and antioxidant use. *Society of Petroleum Engineers Journal* 901–912.

[51] Al-Saadi, F.S., Amri, B.A., Nofli, S. et al. 2012. Polymer flooding in a large field in south oman—initial results and future plans. SPE-154665-MS presented at the SPE EOR Conference at Oil and Gas West Asia, Muscat, Oman, 16–18 April. https://doi.org/10.2118/154665-MS.

[52] Favero, C., Darras, S., and Giovannetti, B. (2013). Process for the enhanced recovery of oil by injection of a polymer solution. World patent WO2013/108174, filed 14 January 2013 and issued 25 July 2013.

[53] Favero, C., Gaillard, N., and Giovannetti, B. (2009). Novel formulations of water-soluble polymers and stabilizing additives for injecting a single compound useable in injection fluids for chemical enhanced oil recovery, Patent WO2010/133258, filed 12 June 2009 and issued 25 November 2010.

[54] Levitt, D.B., Pope, G.A., and Jouenne, S., 2011. Chemical degradation of polyacrylamide polymers under alkaline conditions. Paper SPE 129879 presented at the SPE Improved Oil Recovery Symposium, Tulsa, Oklahoma, USA, 24–28 April.

[55] American Petroleum Institute (1990). *Recommended Practices for Evaluation of Polymers Used in Enhanced Oil Recovery*, API-RP-63. Washington, D.C.: American Petroleum Institute.

[56] Chapman, E.J., Mercer, D., Jerauld, G. et al. (2015). Polymer flooding for EOR in the schiehallion field - porous flow rheological studies of high molecular weight polymers. Paper presented at the 18th European Symposium on Improved Oil Recovery, Dresden, Germany, 14–16 April.

[57] Dwarakanath, V., Dean, R.M., Slaughter, W. et al. (2016). Permeability reduction due to use of liquid polymers and development of remediation options. Paper SPE179657 presented at the SPE Improved Oil Recovery Conference, Tulsa, Oklahoma, USA, 11–13 April.

[58] Jouenne, S., Klimenko, A. and Levitt D. (2016). Tradeoffs between emulsion and powder polymers for EOR. Paper SPE179631 presented at the SPE Improved Oil Recovery Conference, Tulsa, Oklahoma, USA, 11–13 April.

[59] Koh, H. 2015. Experimental investigation of the effect of polymers on residual oil saturation. PhD dissertation. The University of Texas at Austin.

[60] Rubalcava, D. and Al-Azri, N., (2016). Results and interpretation of a high viscous polymer injection test in a south Oman heavy oil field. Paper SPE179814 presented at the SPE EOR Conference at Oil and Gas West Asia, Muscat, Oman, 21–23 March.

[61] Lim, T., Uhl, J.T., and Prud'homme, R.K. (1986). The interpretation of screen factor measurements. *SPE Reservoir Engineering* 1 (03): 272–276.

[62] Zhang, G. and Seright, R.S. 2013. Effect of concentration on HPAM retention in porous media. Paper SPE166265 presented at the SPE Annual Technical Conference and Exhibition, New Orleans, Louisiana, USA, 30 September – 2 October.

[63] ShamsiJazeyi, H., Hirasaki, G.J., and Verduzco, R.. 2013. Sacrificial agents for reducing adsorption of anionic surfactants. Paper SPE164061 presented at the SPE International Symposium on Oilfield Chemistry, The Woodlands, Texas, USA, 8–10 April.

Polymers

[64] Tay, A., Oukhemanou, N., Wartenberg, N. et al. 2015. Adsorption inhibitors: a new route to mitigate adsorption in chemical enhanced oil recovery. Paper SPE174063 presented at the SPE Enhanced Oil Recovery Conference, Kuala Lumpur, Malaysia, 11–13 August.

[65] Wang, D., Cheng, J., and Xia, H. 2001. Viscous-elastic fluids can mobilize oil remaining after water-flood by force parallel to oil-water interface Paper SPE 72123 presented at the SPE Asia Pacific Improved Oil Recovery Conference, Kuala Lumpur, Malaysia, 6–9 October. https://doi.org/10.2118/72123-MS.

[66] Wang, D., Xia, H., Liu, Z. et al. 2001. Study of the mechanism of polymer solution with visco-elastic behavior increasing microscopic oil displacement efficiency and the forming of steady oil thread flow channels Paper SPE 68723 presented at the SPE Asia Pacific Oil and Gas Conference and Exhibition, Jakarta, Indonesia, 17–19 April. https://doi.org/10.2118/68723-MS.

[67] Vermolen, E.C.M., Haasterecht, M.J.T., and Masalmeh, S.K. 2014. A systematic study of the polymer visco-elastic effect on residual oil saturation by core flooding. Paper SPE169681 presented at the SPE Asia Pacific Enhanced Oil Recovery Conference, Kuala Lumpur, Malaysia, 11–13 August. https://doi.org/10.2118/174654-MS.

[68] Urbissinova, T. and Kuru (2010). Effect of elasticity during viscoelastic polymer flooding; a possible mechanism for increasing the sweep efficiency. *Journal of Canadian Petroleum Technology* (also SPE 133471) 49–46. https://doi.org/10.2118/133471-PA.

[69] Delshad, M., Kim, D.H., Magbagbeola, O.A. et al. 2008. Mechanistic interpretation and utilization of viscoelastic behavior of polymer solutions for improved polymer flood efficiency. Paper SPE 113620 presented at the SPE Symposium on Improved Oil Recovery, Tulsa, Oklahoma, USA, 20–23 April.

[70] Hincapie R.E., Rock A., Wegner J. et al. 2017. Oil mobilization by viscoleastic flow instabilities effects during polymer eor: a pore-scale visualization approach Paper SPE185489 presented at the SPE Latin Maerica and Caribbean Petroleum Engineering Conference, Buenos Aires, Argentina, 18–19 May.

[71] Mitchell, J., Lyons, K., Howe, A.M., and Clarke, A. (2016). Viscoelastic polymer flows and elastic turbulence in three-dimensional porous structures. *Soft Matter* 12: 460.

[72] Qi, P., Ehrenfried, D.H., Koh H. et al. 2016. Reduction of residual oil saturation in sandstone cores using viscoleastic polymers. Paper SPE

155.

179689 presented at the SPE Improved Oil Recovery Conference, Tulsa, Oklahoma, USA, 11–13 April.

[73] Howe, A.M., Clarke, A., and Giernalczyk, D. (2015). Flow of concentrated viscoelastic polymer solutions in porous media: effect of mw and concentration on elastic turbulence onset in various geometries. *Soft Matter* 11: 6419. 13. Industrial & Engineering Chemistry Sept 1951, p. 2119.

[74] SNF test F134. OECD 401/GLP, Dec. 1991.

[75] Mallevialle, J., Brucher, A., and Fiessinger, F. (1984). How safe are organic polymers in water treatment? *Journal American Water Works Association* 87–93.

[76] McCollister, D.D., Hake, C.L., Sadek, S.E., and Rowe, V.K. (1965). Toxicologic investigation of polyacrylamides. *Toxicology and Applied Pharmacology* 7 (5): 639–651.

[77] SNF test F242. OECD 203/GLP/Dec. 1995.

[78] SNF test F243. OECD 202/GLP/Dec. 1995.

[79] SNF test F244. OECD 201/GLP/Dec. 1995.

[80] SNF test F245. OECD 301F, DIN38412–27, ISO 7027, GLP/Dec. 1995.

[81] SNF test 0911–01: EC Method A8 (2008) & OECD 107 (1995)/GLP/ Nov. 2009.

[82] Wen, Q., Chen, Z., Zhao, Y. et al. (2010). Biodegradation of polyacrylamide by bacteria isolated from activated sludge and oil-contaminated soil. *Journal of Hazardous Material* 175 (1–3): 955–959.

[83] Wei, L., Rulin, L., Fenglai, L. et al. (2004). Study on the characteristics of a strain of polyacrylamide-degrading bacteria. *Acta Scientiae Circumstantiae* 24 (6): 1116–1121.

[84] Min, Z., Xinyu, W., and Song, H. (2008). Isolation and identification of a hydrolyzed polyacrylamide degrading bacteria strain with Sulfate reducing and study on the characteristics of degrading. *Heiliongjiang Medicine Journal* 21 (6).

[85] Bao, M., Chen, Q., Li, Y., and Jiang, G. (2010). Biodegradation of partially hydrolyzed polyacrylamide by bacteria isolated from production water after polymer flooding in an oil field. *Journal of Hazardous Material* 184: 105–110.

[86] Mutai, B., Jie, P., and Qingguo, C. (2011). Research Progress of biodegradation of polyacrylamide by microorganisms. *Chemical Industry and Engineering Progress* 30: 9.

[87] Hao, C., Liu, Y.J., Wang, D.W. et al. (2008). Studies on degradability of polyacrylamides by a Tetradic bacterial strain mixture and biochemical mechanisms involved. *Oilfield Chemistry* 25 (2): 020.

[88] Yu, F., Fu, R., Xie, Y., and Chen, W. (2015). Isolation and characterization of polyacrylamide-degrading bacteria from dewatered sludge. *International Journal of Environmental Research and Public Health* 12: 4214–4230.

[89] Xia, Y.Y. and She, Y.U. (2007). Optimization of degradation conditions for PAM by seven kinds of bacteria. *Chemistry & Bioengineering* 24: 3.

[90] Joshi, S.J. and Abed, M.M.R. (2017). Biodegradation of polyacrylamide and its derivatives. *Environmental Processes* https://doi.org/10.1007/s40710-017-0224-0.

[91] Guezennec, A.G., Michel, C., Bru, K. et al. (2014). Transfer and degradation of polyacrylamide-based flocculants in hydrosystems: a review. *Environmental Science and Pollution Research*. ISSN: 0944-1344. https://doi.org/10.1007/s11356-014-3556-6.

[92] Caulfield, M.J., Qiao, G.G., and Solomon, D.H. (2002). Some aspects of the properties and degradation of polyacrylamides. *Chemical Reviews* 102: 3067–3083.

[93] El-Mamouni, R., Frigon, J.-C., Hawari, J. et al. (2002). Combining photolysis and bioprocesses for mineralization of high molecular weight polyacrylamides. *Biodegradation* 13: 221–227. Kluwer Academic Publishers.

[94] Charoenpanich, J. (2013). Removal of acrylamide by microorganisms in environmental sciences. In: *Applied Bioremediation – Active and Passive Approaches*. Chapter 5 (ed. Y.B. Patil and P. Rao). ISBN: 978-953-51-1200-6.

[95] Reber, C.A., Khanna, S.N., and Ottenbrite, R. (2007). Thermodynamic stability of polyacrylamide and poly(N,N-dimethyl acrylamide). *Polymers for Advanced Technologies*, Wiley InterScience press. https://doi.org10.1002/pat.949.

[96] Levitt, D.B., 2009. The optimal use of enhanced oil recovery polymers under hostile conditions. PhD dissertation, The University of Texas at Austin.

Chapter five

Polymer Flooding

—

Pilot Design

In this chapter, general guidelines for designing a polymer injection pilot will be provided. Best practices will also be discussed to make the most out of any injection trial.

Essentials of Polymer Flooding Technique, First Edition. Antoine Thomas.
© 2019 John Wiley & Sons Ltd. Published 2019 by John Wiley & Sons Ltd.

5.1. Reservoir Screening – Reminder

Given the current recovery factors (averaging 35% oil originally in place (OOIP) after waterflood), there is a high potential for enhanced oil recovery (EOR) in brownfields, using existing infrastructure to facilitate the implementation. General guidelines were provided in Chapter 4 to assess the feasibility of polymer injection in a field. The two principal screening rules for polymer flooding are:

- Pointing out reservoirs that have poor sweep efficiency due to high oil viscosity and/or large-scale heterogeneity.

- Determining whether the overall conditions are suitable (i.e. compatible brine, mobile oil saturation, retention) for polymer flooding implementation to fix the problem.

Polymer flooding has been applied in both sandstone and carbonate reservoirs. Because injection in carbonates requires a good reservoir understanding and thorough laboratory studies to find the most efficient chemistry, only sandstone reservoirs will be considered here; however, the main screening parameters apply to carbonates as well.

We can narrow the primary parameters needed to check whether polymer flooding is a viable option, by order of importance:

Table 0

Parameter	Preferred condition
Lithology	Sandstones preferred
Wettability	Water-wet
Current oil saturation	Above residual oil saturation
Porosity type (matrix/ fractures)	Matrix preferred
Gas cap	See comments
Aquifer	Edge aquifer tolerated

...

Parameter	Preferred condition
Salinity/hardness	See comments
Dykstra-Parsons and facies variations	$0.1 < DP < 0.8$
Clays	Low (see comments)
Water-cut	See comments
Flooding pattern and spacing	Confined – small spacing

The presence of clays can result in detrimental polymer loss within the reservoir, but solutions exist to minimize their impact on polymer efficiency. The potentially high salinity of the injection water is not a show-stopper but will obviously impact the choice of chemistry and the dosage required to reach the target viscosity.

If screened reservoirs meet these criteria, there is a good chance that polymer flooding is technically viable. The next question is, is it economically viable? This aspect is very dependent on the particular field and should be discussed case by case.

5.2. Pilot Design

The first step to de-risk the technology and evaluate the possible benefits is a pilot test. The primary goals of this approach are the following:

- Prove the technology/concept.
- Obtain valuable information about injection rates, pressure, optimum viscosity, etc.
- Check quality control procedures.
- Have operations personnel work directly with the technology and equipment to better understand the full-scale requirements.

- Check the logistics, delivery, and supply chain.
- Use the information to update the reservoir model.
- Assess the economics of the project.
- Evaluate the produced water treatment technologies.

A pilot will not necessarily break even and should not be designed as such. Proving a concept sometimes requires an innovative approach to de-risk as many parameters as possible and find the optimum injection strategy for full-field deployment (Figure 5.1).

5.2.1. Pattern Selection

Once a reservoir with acceptable characteristics has been chosen and polymer flooding is deemed applicable, the design can start. The first step consists of selecting the most appropriate zone for pilot injection. This decision is based on two main parameters:

- Finding a zone that is relatively representative of the entire reservoir.
- Minimizing the response time to obtain valuable information, in order to quickly decide on long-lead items and begin the approval process for a full-field expansion.

The response time varies based on many factors including spacing, reservoir thickness, injection rates, production history, etc. The general ideas are as follows:

- Select a confined pattern where the oil production from polymer injection can be isolated. For example: for vertical wells, five spots with a central producer; and for horizontal wells, two injection wells and one producer.
- Optimize the spacing to maximize efficiency. For verticals, a spacing between 100 and 150 m is preferred. For horizontals, a length of 1000 m maximum and a spacing of 100 m

Polymer flooding:
from design to implementation

Figure 5.1

 Preliminary screening

 2 weeks

- Gather reservoir description (including temperature, salinity, and permeability for polymer selection).
- Compare to other fields in literature.
- Select potential polymer candidates.

2 Preliminary laboratory tests

1 to 3 months

- Polymer rheology and shear sensitivity.
- Determination of target concentration/viscosity.
- Compatibility tests with other chemicals.
- Basic simulation tests.
- Basic economic analysis.

3 Detailed laboratory tests
Equipment design

6 to 8 months

- Long-term stability tests and coreflooding.
- Improved reservoir simulation.
- Field test and equipment design / feasibility study.
- Polymer selection.

4 **Field test and pilot**

 6 to 18 months

- Test polymer mixing facilities.
- Assess injectivity and maximum rates and viscosity.
- Evaluate oil recovery.
- Assess back-produced water treatment.
- Update reservoir model and economic analysis.

5 **Full field deployment**

1 to 20 years

- Dissolution and injection facilities deployment.
- Logistics.
- Monitoring / Surveillance plan.
- Update reservoir model.

are suitable options. If the distance is less, early polymer breakthrough may impact efficiency.

- Check the connectivity between the wells (tracer tests, pressure tests, production history, and so on).

- If the reservoir is multilayered, if possible, isolate a zone for injection.

- Select a zone far from the water-oil contact. If this is not possible, study the possibility of recompleting the well and isolating the zone.

- Check well completion and cleanliness. For cased vertical wells, a minimum of 12 perforations per foot is needed to minimize shear through the completion. Big bore (diameter and length) perforations can be used to further increase surface area for injection and reduce sandface shear. Acid jobs can be performed to clean the wells before the start of injection. For horizontals, perforated liners or wedge-wrapped screens can be used with minimum degradation.

A detailed pattern analysis can be continued once the number of candidates has been narrowed using these criteria (Figure 5.2).

5.2.2. How Much Polymer?

Defining a target viscosity (we should even say a resistance factor) is usually a prerequisite before starting any laboratory study or pilot design. If the reservoir is heterogeneous, with crossflow between layers, a calculation using Darcy's law shows that the ideal viscosity is equal to

$$\mu_{polymer} = \textbf{mobility ratio} * \textbf{permeability contrast}$$

Five-spot and inverted five-spot patterns

Figure 5.2

- Chemical injector
- Water injector
- Producer

The permeability contrast can be simply defined as the higher permeability divided by the lower permeability for adjacent layers with crossflow. In a case where no crossflow occurs, this term can be removed from the equation and only the mobility ratio considered. Seright [1] details the Daqing case where the end-point mobility ratio is 10 and the permeability contrast is 4, yielding an optimum polymer viscosity of 40 cP. Obviously, in some heavy oil pools, this strategy is barely applicable, given the viscosity needed and the associated costs (not including injectivity aspects). But when possible, the mobility ratio value targeted should be less than 1, to take heterogeneities into account [2].

The volume of polymer injected is also very important. A rule of thumb is that at least 30% (or 50% in heavy oil pools) of the reservoir pore volume should be filled with polymer. However, from a reservoir engineering standpoint, and for good efficiency, the more the better. In Daqing, 60% of the reservoir has been filled with polymer; the volume is 80% in Mangala oil field (India), 50% in Shengli oil field (China), and 60% in Suffield (Canada) [1]. Stopping the injection depends largely on economics. When the cost of injecting polymer exceeds the benefits coming from the oil production, the process should cease [1], or a slug with a viscosity taper can be implemented to protect the back end of the entire polymer slug. A producer can be shut in when the water-cut increases again to uneconomical values. If only 30% of pore volume has been injected, but oil is produced at economical rates, polymer injection should be continued. From a technical standpoint:

- A small polymer bank will not be efficient for recovering bypassed oil. When switching back to water, the water will not push the polymer slug homogeneously toward the producer but rather will finger through it in the high-permeability zones, where the residual resistance factor is lowest. The consequences can even be worse if crossflow occurs.

- A large polymer bank can compensate for retention and maintain viscosity during the entire transit through the reservoir.

Additionally, in heterogeneous reservoirs, the size of the bank injected will have a huge impact on the following:

- *The choice of the polymer.* In theory, the polymer will be effective – and therefore should be stable – only over the portion of the reservoir injected. As stated earlier, when switching back to water, it will finger through the slug and dilute it. This should help determine the optimum chemistry from both a technical and an economic standpoint.

- *The concentration of polymer that can be expected at the production side.* If less than 1 pore volume is injected, the dilution effect caused by water injection and polymer retention in the virgin parts of the reservoir will reduce polymer concentration. Therefore, it is highly unlikely that the original polymer concentration will ever be seen in production wells. In cases where rapid tracer transit profiles exist, where fractures from injector and producer may be connected, or in close-spacing horizontal injector/producer pairs with severe channeling, early polymer breakthroughs of high concentration can occur. The immediate result may be that the well must be shut in and that polymer injection can be continued only after a conformance treatment. If the produced polymer is being diluted by large produced brine volumes, it may not be an issue; however, the ability to produce to temporary tanks may minimize production facility upsets if they occur. This should be understood when designing the water treatment facilities and when health, safety, and environment (HSE) concerns are raised for offshore implementation.

Simulation studies and sensitivity analyses can be performed to optimize bank size, determine the impact of water chase on slug integrity, and evaluate the benefits of injecting a higher-viscosity polymer plug.

Once the parameters have been set, it is possible to design the injection protocol and set a list of variables that should be monitored during the injection.

5.2.3. Injection Protocol

5.2.3.1. Start-Up of Polymer Injection

Before the start of injection, the injection wells should be cleaned to allow good injectivity. When injecting in a mature reservoir, it is paramount to have a good water injection baseline, with stable injection rates and pressures to compare with polymer injection. There are different ways to start a polymer injection:

1. Begin with the same injection rate as waterflood, and progressively increase the viscosity while monitoring reservoir response: for instance, one-third of target viscosity, then two-thirds, and finally target viscosity.

2. Begin with a target viscosity of one-third the target injection rate, then two-thirds, and finally the full injection rate.

The first strategy helps obtain a good understanding of the flow regime, differentiating between linear and radial/matrix flow. In theory, if the flow is radial, an increase in viscosity from 1 to 2 cP at a given rate should lead to almost 50% injectivity loss (compared to water; see Section 5.3). If all injection guidelines to prevent polymer degradation have been respected, and if no significant injectivity loss is observed, then microfractures are likely present (see Section 5.3).

Hall plots can be used to monitor the reservoir response (see Section 5.4). If pressure does not stabilize and reservoir integrity is threatened, the best strategy is to decrease the injection rate without adjusting the viscosity.

5.2.3.2. Ending Polymer Injection

No clear rule defines how to stop polymer injection, but, generally speaking, two approaches are possible:

1. Polymer injection is simply stopped, and water injection is resumed.

2. Polymer viscosity is ramped down over a period of time to minimize mobility contrast between the polymer slug and chase water. In theory, this should help slightly delay water breaking through the slug and reaching the producers.

However, in both cases, once water injection is resumed, there is a high risk that the water-cut will rise rapidly in the producers, along with small fractions of the injected polymer, which can occur over a very long period.

5.2.3.3. Voidage Replacement Ratio (VRR)

The voidage replacement ratio (VRR) is the ratio of reservoir barrels of injected fluid to reservoir barrels of produced fluid. Balancing injection and production is a common reservoir management practice for conventional oil-bearing formations. Recent studies [3, 4] and analyses have suggested, for heavy oils, that targeting a VRR below 1 could improve recovery by:

- Allowing compaction in compressible formations.
- Decreasing reservoir pressure and using a secondary gas-cap drive.

Other authors have discussed this topic and shown that no clear correlations could be made with current field data, especially for Pelican Lake, where comprehensive information is available [5]. The compaction/dilation effect was also studied for a specific field in Suriname, showing that the injection strategy was indeed important, in addition to reservoir characteristics (heterogeneity), to explain the results observed in the field. The conclusions of the work were that there is an optimum rate, pressure, and viscosity to flood compressible formations. High injection rates could result in pressure and energy waste, dilating the formation and impairing oil displacement [6]. Other possible issues with VRR < 1 include the following:

- If more fluid is drawn, it will be difficult to pressurize the reservoir, leaving injection wells under vacuum in some cases. Pressure is therefore not a good indicator of the efficiency of polymer injection.

- Water fingers are likely to form when VRR < 1, bypassing large volumes of oil. When switching to polymer injection, given the pressure gradient, the slug might enter these zones preferentially, with deteriorated efficiency and, possibly, early polymer breakthrough.

In conclusion, the injection strategy depends on the reservoir and field constraints. A thorough reservoir analysis is required to understand all the parameters that influence oil recovery. For polymer injection, it is recommended, when possible, to balance injection and production; this will help with understanding the changes in fluid production and will enhance history matching.

A reservoir management strategy to maximize oil recovery may consist of purposefully building reservoir pressure at the start of polymer injection. For example, a strategy for a simple newly drilled pattern with two horizontal injectors and a central producer could be as follows:

- Shut in the producer.

- Inject a viscous polymer solution in both injection wells, quickly building pressure until the maximum allowable pressure is reached, indicating reservoir fill-up. This pressure is usually mandated by the governments or determined from the reservoir fracture gradient and is monitored to not compromise reservoir and cap rock integrity.

- Open the producer.

- Decrease the injection rate slightly, keeping relatively low injection rates to minimize polymer breakthrough.

This strategy has the advantage of building a homogeneous polymer front in the injection well, maximizing the reservoir length contacted and minimizing the formation of water/polymer fingers. An obvious drawback is the absence of fluid production during pressure build-up. This time can be shortened by optimizing the spacing and injection strategy.

5.3. Injectivity

Concerns about injectivity issues are often raised, since a viscous solution is injected instead of water [7]. Mathematically speaking, the injectivity index, II, is defined as:

$$II = \frac{Q}{P_{bhi} - P_e} = \frac{k_w \cdot h_i}{141.2 \cdot \mu_w \cdot B_w \cdot \left(\ln \frac{r_e}{r_w} + S \right)}$$

where Q = injection rate (std bbl/d); P_{bhi} = bottom-hole pressure; P_e = reservoir pressure (in psi); k_w = permeability (mD); μ_w = water viscosity (cp); B_w = water formation volume factor (res vol/ST vol) r_w and r_e are the wellbore and drainage radii, respectively (ft); h_i = injection height (ft); and S = total near-wellbore skin.

Looking at this equation, there are two important parameters for polymer injection: bottom-hole pressure and viscosity. It is very difficult to know these values precisely at any point in the injection facilities, especially in the well and the near-wellbore area, for these reasons:

- Polymer solutions are non-Newtonian fluids, so viscosity changes with the shear rate applied. In the well, where the shear is relatively high, viscosity will be low, which will also have an impact on the pressure drop during flow.

- Polymers act as friction reducers [8]. For the same injection rate, the pressure drop in the pipes when injecting polymer may be 70% less than with water. Therefore, predicting the pressure drop in the pipes and the pressure at the bottom-hole will be difficult. Care should be taken when attempting to run a 3D model to predict injectivity (Figure 5.3).

A clarification is required regarding the definition of an injectivity issue. Given the previous equation and polymer injection,

Figure 5.3 **Pressure drop average vs. Reynolds number in a 4" pipe. Comparison between water and a polymer solution @10cP at 7,34 s^{-1}**

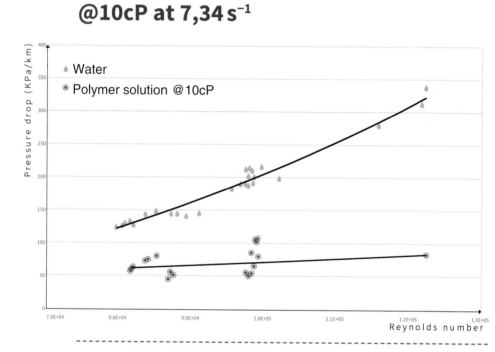

an injectivity issue it translates into a decrease of the injected volume into a given reservoir, compared to waterflood. Let's consider three cases:

1. Injectivity decreases dramatically right away upon commencement of polymer injection. There could be several reasons, including water quality, a damaged wellbore, or an abrupt change in fluid flow due to viscosity. In any case, the first response is to decrease the injection rate, keeping the same viscosity, to see if the pressure stabilizes. If the pressure does not stabilize, polymer injection should cease, the reasons should be investigated, and the well should be cleaned with

- Water only, at first. If the pressure doesn't decrease, then …

- Add an oxidizer or a chemical not compatible with the polymer, to break/degrade the molecule (sodium persulfate, bleach, or tetrakis hydroxymethyl phosphonium sulfate [THPS]). Conventional acid will not remove the polymer. If pressure still doesn't decrease (and the polymer is not at fault), then it is likely that solids, oil carry-over, or a combination with the polymer is the cause of the observed injectivity decrease. At that stage, conventional acids can be considered.

- If the damage is too severe (fine migration, internal filter cake), fracturing may be required to restore injectivity.

2. Injectivity decreases after a time. This should normally occur if the polymer slug mobilizes an oil bank: in that case, the pressure will slowly build up, possibly forcing a decrease in injection rate depending on reservoir and facility constraints. However, when the VRR is less than 1, pressure buildup in the reservoir may take a very long time and, in some cases, may not happen at all.

3. Injectivity barely changes or increases. Several explanations are possible:

- Friction reduction effect, changing the pressure drop and mathematically affecting the injectivity index.
- Fracture extension and/or creation.
- Increase in the surface/area swept. If the flooding front enlarges, there is less pressure drop per linear meter.
- Viscosity loss due to degradation. Increasing the concentration temporarily will increase the viscosity, which can help detect changes. Also, variations in water-cut or oil cuts can prove that viscosity remains in the reservoir, is sweeping more than before, and is mobilizing an oil bank.
- Injection out of zone (gas cap or water aquifer) and other issues related to reservoir management.

Looking at all existing and past projects, published or not, very few cases have been reported where an injectivity decrease was observed at the start of polymer injection. A probable reason is the presence of microfractures in the near-wellbore area created during the well's drilling/completion or during water injection (or injection of cold water into a hot reservoir). Some existing flow paths are beneficial to polymer injection, in that microfractures can be extended slightly, decreasing the shear rate in the near-wellbore area and therefore minimizing possible mechanical degradation [9, 10]. In some specific cases, it is also possible to pre-shear the polymer solution to remove very-high-molecular-weight molecules and enhance injectivity. However, some benefits may be lost, such as viscoelasticity. One last method to improve polymer injectivity is hot-polymer injection, whereby the polymer is warmed prior to injection and viscosity in the near-wellbore area is low; as the polymer flows further from the wellbore, it is cooled in the reservoir, gaining in situ viscosity.

5.3.1. Discussion on Injectivity

Understanding what happens in the near-wellbore area is critical to the success of polymer injection. To build a business case, it is necessary to make predictions about the volumes injected and assume that the viscosity will remain intact during the project. In the near-wellbore area, the main risk is mechanical degradation, either through perforations or at the sand face.

First, a differentiation should be made between vertical and horizontal injectors. Very often, the characteristics of horizontal wells make them good candidates for polymer injection: their length is important and the injection rate is not very high, especially onshore in shallow reservoirs. The area available for flow is therefore important, with low shear rates and little chance of mechanical degradation. Authors have also discussed the impact of completion [11], compaction/dilation, and fracture propagation in unconsolidated formations [12]. It is also possible to calculate the shear through orifices for completion with screens.

For verticals, completion is critical to ensure that little to no degradation occurs. The length and diameter of the perforation can be maximized to reduce shear; and the more perforations, the better (>12 shots per foot). But understanding the flow regime is also critical: the pressure drop and shear rate will not be the same if fractures exist.

Well tests and basic monitoring can help understand what happens in the well. First, it is necessary to obtain a steady water-injection baseline and test reservoir pressure limits. Hall plots combined with pressure tests and other monitoring tools (step rate test) should be used to determine the parting pressure and set a maximum allowable pressure limit not to cross to minimize fracture propagation.

Another simple tool consists of analyzing injectivity through the value q/ΔP:

$$\text{If } \frac{q}{\Delta P} \leq \frac{\Sigma kh}{141.2\,\mu\,ln\left(\frac{r_e}{r_w}\right)}, \text{ radial flow is probable.}$$

$$\text{If } \frac{q}{\Delta P} \geq \frac{\Sigma kh}{141.2\,\mu\,ln\left(\frac{r_e}{r_w}\right)}, \text{ linear flow is probable.}$$

Where q = injection rate (barrels per day [BPD]), P = pressure drawdown (psi), k = permeability (md), h = formation height (ft), μ = fluid viscosity (cp), r_e = external drainage radius (ft), and r_w = wellbore radius (ft).

In order for this relationship to prove useful, very accurate measurements of the bottom-hole injection and reservoir pressures should be known. In addition, flow capacity can be best represented if the injection well in question has been cored.

A careful review of existing field cases shows that most of the injections (water and polymer) were done under fracturing conditions [9, 10, 13–15]. Several reasons can be given;

- Drilling a well changes the stresses around the bore, damaging the formation and weak zones.

- For cased, cemented, and perforated vertical wells, the perforation creates a fracture. The injected fluids will simply make the fracture propagate over time.

- In some cases, cold water injected into hot reservoirs can initiate thermal fractures.

Given these possibilities, there is little chance that polymer will encounter the rock matrix immediately after the perforations.

The best way to check injectivity is to run a trial with polymer: injecting a relatively low viscosity will give information about the flow regime. In theory, if the flow is radial, the following equation can be used to predict injectivity, relatively to water:

$$\frac{I}{I_0} = \frac{ln\left(\frac{r_e}{r_w}\right)}{\left[F_r \, ln\left(\frac{r_p}{r_w}\right) + ln\left(\frac{r_e}{r_p}\right)\right]}$$

with I/I_0 = injectivity relative to water; r_p = radius of polymer front (ft); r_e = external drainage radius (ft); r_w (wellbore radius); and F_r = resistance factor.

In that case, given an external drainage radius of 1000 ft, a wellbore radius of 0.3 ft and a resistance factor for polymer of 2 (1 for water), we obtain the plot of I/I_0 versus radius of polymer front shown in Figure 5.4.

Assuming no fractures, injectivity relative to water should decrease by 50% just by increasing the resistance factor from 1 to 2. First, this has never been observed in any field case documented. Second, if, during the trial, such a decrease is not observed, then it is likely the flow is not purely radial and microfractures exist. Such a test can be performed at a low injection rate to create favorable conditions for the polymer, to minimize potential mechanical degradation.

If fractures exist, there are important consequences:

- The pressure drop – and therefore shear – are dramatically decreased in linear flow, limiting polymer mechanical degradation [1].

- The shear-thickening behavior disappears. Shear-thickening is a phenomenon observed in the laboratory at very high flow rates in a core (matrix flow). When polymer molecules

Injectivity relative to water, assuming no fractures (external drainage radius = 1000 ft, wellbore radius = 0.3 ft; resistance factor = 2)

Figure 5.4

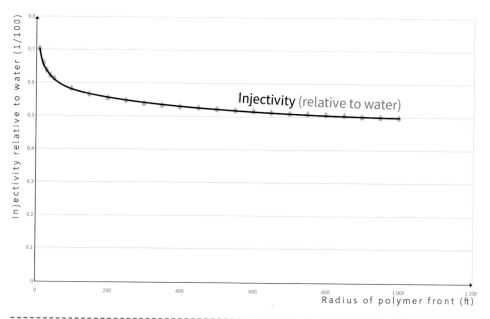

are forced through pores at high shear, the resistance factor greatly increases, to a point that mechanical degradation occurs. In the field, if fractures are present (no matrix flow), the surface opened to flow is much higher, and the polymer molecules encounter less resistance. Moreover, as shown in Figure 5.5, shear decreases very quickly once the fluid enters the reservoir, rapidly leaving the region where shear-thickening could have occurred. Correlating fracture length with shear rate in the near-wellbore area is therefore a good way to de-risk mechanical degradation.

Using microfractures or forcing their creation will help minimize mechanical degradation and maintain reasonable injection rates. Obviously, their extension should be controlled such that there is no direct connection between the injector and

Figure 5.5 | Shear vs. distance from the wellbore

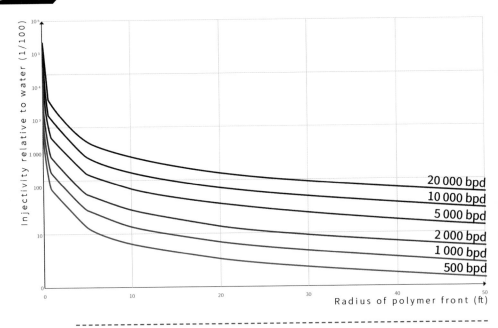

producer; the maximum distance should not be equal to more than one-third the spacing. Fracture propagation can be followed with pressure and tracer tests, for instance. If rapid connection is observed, the injection rate should be decreased to allow partial closure of the fracture.

A thorough geomechanical analysis is also a good tool to understand fracture propagation and directions (vertical, horizontal, new versus reopened): geological properties of the reservoir, layering, facies, reservoir production history, reservoir depth, and injection rates will often limit fracture propagation (fluid leak-off). Considerable information can be gathered from the hydraulic fracturing industry, where many efforts have been undertaken to understand fracture initiation and growth in tight formations. Below a certain depth, fractures grow vertically. Their growth will be limited by other geological formations or barriers, and there will be an increasing loss of fluid as it leaks into the most permeable zones [16]. In many formations, fracturing can be envisioned as similar to shattered glass, creating a complex fracture network that follows heterogeneities and whose extension is rapidly limited by fluid leak-off (Figure 5.6).

Changes at the production side are another aspect that should be considered when discussing injectivity and injection rates. If the injection rate is decreased, but, at the production side, the oil-cut and water-cut percentages are switched in favor of oil production, then an economic balance can be reached. Here is a simple example: assume a void replacement ratio of 1 with $100\,m^3\,d^{-1}$ water injected and $100\,m^3\,d^{-1}$ of total fluid produced. If the water-cut is 95% during the waterflood, then only $5\,m^3\,d^{-1}$ of oil are produced. Now assume a decrease of 25% injectivity during polymer injection, and keep a void replacement ratio of 1; the total fluid injected and produced equal $75\,m^3\,d^{-1}$. Assuming a decrease in water-cut of 10% down to 85%, this translates to oil production of more than $11\,m^3\,d^{-1}$, doubling the rate during waterflood. The fact that water-cut decreases has a significant economic impact, since less water has to be treated and/or disposed of.

Hall plot – general description

Figure 5.6

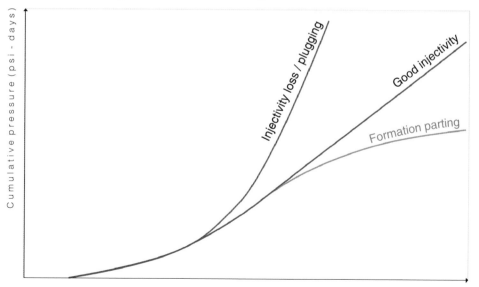

5.4. Monitoring

Before polymer injection, several tests can be performed to gain a better understanding of the reservoir and near-wellbore area. These tests are also paramount for establishing a good baseline before switching to polymer injection. Typical tests include the following:

- Step rate tests (to determine parting pressure).
- Pressure fall-off and pressure build-up (injectivity/productivity, effective reservoir properties, and in situ polymer viscosity).
- Pulse tests (to check well connectivity).
- Production logging tool (PLT) (to identify the main reservoir units and reservoir conformance).
- Tracer tests (to check well connectivity and identify heterogeneities).
- Saturation logs (to ensure enough oil is left; also useful in surfactant-polymer [SP] and alkali-surfactant-polymer [ASP] design).

Water quality can be monitored in conjunction with probes for oxygen and salinity (conductivity), and periodic sampling for iron content. The level of contaminants should be clearly determined to adapt the injection strategy, water treatment facilities, and choice of polymer and its protective package (if required).

During polymer injection, pressure and injection rates are usually continuously recorded at the injection side. The Hall plot, shown in Figure 5.6, can be used to track trends and detect either plugging or fracture formation. At the production side, parameters such as total fluid produced, water-cut, and oil-cut should be recorded. In the long term, for a brownfield, we should look at the following:

- Peak oil rate, breadth of peak, and response time to flood.
- Average oil rate and oil cut after peak response.

- Average sustained oil rate and oil cut to current date of flood.
- Average early injection rates.
- Sustained injection rates.
- Time before water breakthrough.
- Total oil recovered during the pilot.

In theory, the results of a pilot (total oil recovered) should be analyzed once the water-cut again reaches the economic limit (98% or 99%).

5.5. Modeling

3D modeling is a useful tool that can help predict the efficiency of polymer flooding, assuming the user has a good understanding of polymer physics and flow properties. The most critical part is modeling the near-wellbore area to duplicate the results observed in the field during pilot injection. Two runs should be made and their differences analyzed:

- Injection with a fixed bottom-hole pressure. Usually, the injection rate will be limited quickly by the simulator because of increased viscosity, even though this is often not observed in the field.

- Injection with a fixed rate, close to the waterflood rate. This run will probably be closer to the actual field data, given the rheology of the solution and near-wellbore features that usually are not accounted for (fractures, drag reduction, improved sweep efficiency, etc.).

Keywords related to shear-thinning or shear-thickening properties should be used carefully, with a good understanding of the wells' completion and the presence of damages or fractures. The most important parameters that can be obtained from laboratory experiments are resistance factors and retention. If no porous-media experiment has been performed, it is possible to use the low-shear viscosity obtained in conventional rheometers for simulation purposes, since, in the middle of the reservoir, the polymer solution will likely show Newtonian behavior given the low shear rates. Polymer degradation in the reservoir is usually limited, since the chemistry is always selected to remain stable during propagation throughout the reservoir. Input can be gathered from long-term stability tests. A conservative approach should be taken for the residual resistance factor, using a value of 1. Sensitivity analyses should be made of slug size and viscosity, based on the water chase and a low R_k value, as mentioned earlier.

The model should be updated during the pilot injection and the results used to build the true business case.

5.6. Quality Control

The most important parameter in polymer flooding remains viscosity. Not everything can be controlled or anticipated, but many precautions can be taken to minimize the risk of degradation. If all engineering guidelines (for both surface facilities and well completions) are respected, along with close monitoring of water quality, then chances are that the viscosity will remain intact when entering the reservoir. There are currently two ways to control viscosity at the wellhead:

• Manual sampling

• Inline viscosity monitoring

If not carried out properly, manual sampling can lead to polymer degradation due to the following:

• Oxygen ingress during sampling and/or viscosity measurement.

• Mechanical degradation resulting from the way the sample is taken or the sampling device itself, i.e. excess pressure drop across a valve or orifice.

Specific sampling methods and devices have been developed to avoid degradation during sampling and the measurement itself. The main drawback of manual sampling is manpower. To solve this problem, inline viscometers have been developed that display the viscosity value in real time [17]. Discrepancies from expected viscosity values can be observed quickly and actions taken to remedy the problem.

Product quality can be controlled with standard tests such as yield-viscosity determination and filter ratio. All indications are provided in the certificate of analysis along with procedures to verify that the product falls within the specifications.

5.7. Specific Considerations for Offshore Implementation

The design of polymer injection offshore is complex for several reasons. The first is footprint. When infrastructure is already in place, space and weight limitations often make the installation of new equipment complicated. This aspect, together with local constraints (weather and marine conditions), will dictate the choice of product form (powder or emulsion). The second aspect is related to the presence of devices such as chokes, which are difficult to replace or change and can dramatically degrade the polymer solution. Solutions exist that can still make the process viable:

- *Chemical solutions.* Adapt the polymer chemistry and molecular weight, or develop delayed-action polymers that will uncoil once in the reservoir.

- *Mechanical solutions.* Non-shearing chokes can replace the existing ones or be part of greenfield development.

A third aspect that impacts polymer injection is well spacing. Offshore, the residence time in the reservoir is quite long, and the response time is a matter of years. Without a decrease in spacing to get faster results, little can be done to accelerate the process:

- EOR should be started on day 1 to minimize fingering and unwanted water production.

- Viscosity can be increased to play on the conformance effect and also limit retention and degradation and improve sweep while delaying polymer breakthrough.

- A large enough pore volume should be injected to offset retention and delay water breakthrough during the final water chase.

The fact that distances are very significant drives people to choose very robust chemistries, always assuming that the polymer will be efficient all the way from the injection well to the producer. With current developments and new chemistries, polymers have robust designs that provide good stability. What might not be stable is the displacement of the polymer slug by the water chase, which will render the process inefficient and the use of expensive chemistries not completely justified. A compromise should be found to minimize expenses while ensuring good sweep efficiency and mobilizing oil in the long term.

Finally, water treatment can be a problem offshore if polymer is present in the effluents and the latter must be disposed of. The polymer will negatively affect standard equipment such as induced gas flotation (IGF), dissolved gas flotation (DGF), and hydrocyclones. The impact will greatly depend on how much polymer is produced and, therefore, how much polymer is injected (see the discussions in previous sections). In all cases, the best (and often only) strategy is to reinject the water containing polymer after separation and treatment. Studies are ongoing to understand the fate of polymer in a marine environment (see Chapter 4).

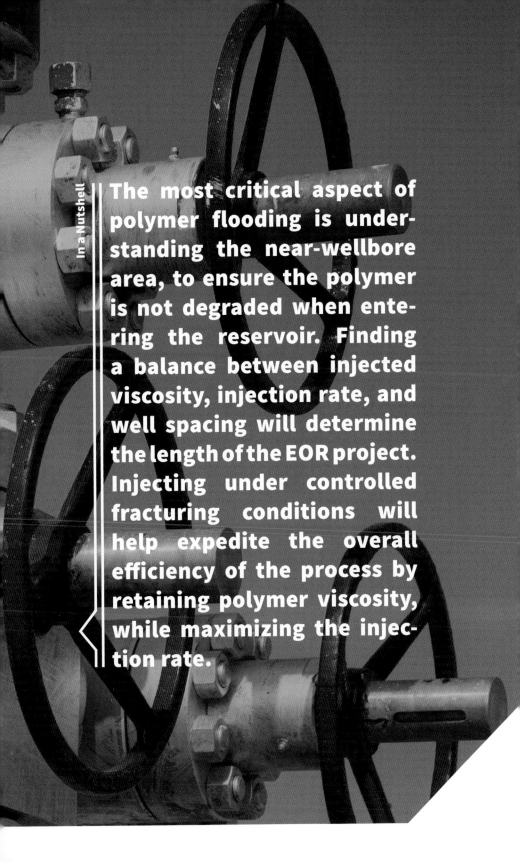

The most critical aspect of polymer flooding is understanding the near-wellbore area, to ensure the polymer is not degraded when entering the reservoir. Finding a balance between injected viscosity, injection rate, and well spacing will determine the length of the EOR project. Injecting under controlled fracturing conditions will help expedite the overall efficiency of the process by retaining polymer viscosity, while maximizing the injection rate.

References

[1] Seright, R.S. (2016). How much polymer should be injected during a polymer flood? Paper SPE 179543 presented at the Improved Oil Recovery Conference, Tulsa, Oklahoma, USA, 11–13 April. https://doi.org/10.2118/179543-MS.

[2] http://www.prrc.nmt.edu/groups/res-sweep/poly-flood-videos.

[3] Vittoratos, E. and Kovscek, A. (2017). Doctrines and realities in reservoir engineering. Paper SPE 185633 presented at the 2017 SPE Western Regional Meeting, Bakersfield, California, USA, 23 April. https://doi.org/10.2118/185633-MS.

[4. Vittoratos, E.S. and West, C.C. (2013). VRR < 1 is optimal for heavy oil waterfloods. Paper SPE 166609 presented at the SPE Offshore Europe Oil & Gas Conference and Exhibition, Aberdeen, Scotland, 3–6 September. https://doi.org/10.2118/166609-MS.

[5] Delamaide, E. (2017). Investigation on the impact of voidage replacement ratio and other parameters on the performances of polymer flood in heavy oil based field data. Paper SPE 185574 presented at the SPE Latin America and Caribbean Petroleum Engineering Conference, Buenos Aires, Argentina, 18–19 May. https://doi.org/10.2118/185574-MS.

[6] Wang, D., Seright, R.S., Moe Soe Let, K.P. et al. (2017). Compaction and dilation effects on polymer flood performance. Paper SPE 185851 presented at the 79th EAGE Conference and Exhibition, Paris, France, 12–15 June. https://doi.org/10.2118/185851-MS.

[7] Glasbergen, G., Wever, D., Keijzer, E. et al. (2015). Injectivity loss in polymer floods: causes, preventions and mitigations Paper SPE 175383 presented at the SPE Kuwait Oil & Gas Show and Conference, Mishref, Kuwait, 11–14 October. https://doi.org/10.2118/175383-MS.

[8] Toms, B.A. (1948). Some observations on the flow of linear polymer solutions through straight pipe at large Reynolds numbers. In: *Proceedings of the International Congress on Rheology, Scheveningen, Netherlands*, vol. 2, 135–141.

[9] Van der Heyden, F.H.J., Mikhaylenko, E., de Reus, A.J. et al. (2017). Injectivity experiences and its surveillance in the West Salym ASP pilot. Paper EAGE ThB07 presented at the 19th European Symposium on Improved Oil Recovery, Stavanger, Norway, 24–27 April.

[10] Spagnuolo, M., Sambiase, M., Masserano, F. et al. (2017). Polymer injection start-up in a brown field - injection performance analysis and subsurface polymer behavior evaluation. Paper EAGE Th B01 presented at the 19th European Symposium on Improved Oil Recovery, Stavanger, Norway, 24–27 April.

[11] Bouts, M. N. and Rijkeboer, M.M. (2014). Design of horizontal polymer injectors requiring conformance and sand control. Paper SPE169722 presented at the SPE EOR Conference at Oil & Gas West Asia, Muscat, Oman, 31 March – 2 April. https://doi.org/10.2118/169722-MS.

[12] Zitha, P.L.J., Barnhoor, A., and Logister, R. (2015). Pressure behavior and fracture development during polymer injection in a heavy oil saturated unconsolidated sand. Paper SPE 174232 presented at the SPE European Formation Damage Conference and Exhibition, Budapest, Hungary, 3–5 June. https://doi.org/10.2118/174232-MS.

[13] Al-Saadi, F.S., Amri, A.B., Nofli, S. et al. (2012). Polymer flooding in a large field in South Oman – initial results and future plans. Paper SPE 154665 presented at the SPE EOR Conference at Oil and Gas West Asia, Muscat, Oman, 16–18 April. https://doi.org/10.2118/0113-0082-JPT.

[14] Anand, A. and Ismali, A. (2016). De-risking polymer flooding of high viscosity oil clastic reservoirs – a polymer trial in Oman. Paper SPE 181582 presented at the SPE Annual Technical Conference and Exhibition, Dubai, UAE, 26–28 September. https://doi.org/10.2118/181582-MS.

[15] Moe Soe Let, K.P., Manichand, R.N., and Seright, R.S. (2012). Polymer flooding a 500cp oil. Paper SPE 154567 presented at the 18th SPE Improved Oil recovery Symposium, Tulsa, Oklahoma, USA, 14–16 April. https://doi.org/10.2118/154567-MS.

[16] King, G.E. (2012). Hydraulic fracturing 101: what every representative, environmentalist, regulator, reporter, investor, university researcher, neighbor and engineer should know about estimating frac risk and improving frac performance in unconventional gas and oil wells. Paper SPE 152596 presented at the SPE Hydraulic Fracturing Technology Conference, The Woodlands, Texas, USA, 6–8 February. https://doi.org/10.2118/0412-0034-JPT.

[17] Bonnier, J., Rivas, C., Gathier, F. et al. (2013). Inline viscosity monitoring of polymer solutions injected in chemical enhanced oil recovery processes. Paper SPE 165249 presented at the SPE Enhanced Oil Recovery Conference, Kuala Lumpur, Malaysia, 2–4 July. htps://doi.org/10.2118/165249-MS.

Chapter six

Engineering

Polymer selection is a critical part of the design of a chemical enhanced oil recovery (EOR) project. Once the right chemistry and molecular weight have been selected, it is necessary to design injection facilities so the macromolecules experience as little modification as possible during the entire journey from the surface down to the reservoir. The previous chapters detailed the main factors affecting polymer stability. Obviously, when designing injection facilities, the main goal is to limit any chemical or mechanical degradation as much as possible. Specific guidelines have been developed, some of which are detailed in this chapter.

Essentials of Polymer Flooding Technique, First Edition. Antoine Thomas.
© 2019 John Wiley & Sons Ltd. Published 2019 by John Wiley & Sons Ltd.

6.1. Preliminary Requirements

6.1.1. Water Quality

Field experience has shown that a well-performing waterflood with clean water and good injectivity will generally translate into a well-performing polymer flood. In any waterflood, the presence of oil carry-over and suspended solids in the injection water can have a detrimental impact on long-term injectivity. The fewer contaminants, the better. The polymer itself is not influenced by the presence of oil or solids, even though flocculation can occur with particles. Oxygen is more problematic both for the polymer (degradation) and the injection facilities (corrosion, bacterial development). Its concentration should therefore be decreased as much as possible. Iron and H_2S may also affect polymer viscosity negatively if combined with oxygen; alone, they have little impact on the viscosity of the solution. However, a high H_2S concentration (greater than several parts per million) will create corrosion problems.

Questions are often raised regarding possible gel formation if crosslinking occurs between the anionic polymer and iron (III), mainly based on laboratory observations. Looking at the stability diagram of iron, it appears that Fe^{3+} as an ion exists as a dominant species only at pH below 2 and very high reduction potential (Eh, above 0.8), making its occurrence in field brines very unlikely. If a redox reaction occurs, iron (II) will be transformed into iron oxides that will not crosslink the polymer but rather will be flocculated by it. The main uncertainty is linked to the transition from iron (II) to iron oxides, during which radicals can be generated and break the polymer backbone. To minimize polymer degradation, any reaction (expected or not) with oxygen should occur before the polymer is dissolved in water.

A summary of typical water specifications is given next. These depend on operator requirements; higher values can be accepted if compatible with the particular reservoir:

- Oil content <100 ppm (to minimize reservoir plugging).

- Solids content: <20 ppm, size <5 μm (to minimize reservoir plugging).

- Consistent salinity over the life of a project, to reduce viscosity variations.

- Oxygen content <40 ppb (below 60 °C) and <10 ppb (above 60 °C).

- Iron or hydrogen sulfide content: depends on the oxygen content in the water (see Chapter 4).

Oxygen content and salinity variations can be checked with inline probes or periodic sampling and analysis. It is difficult with conventional tests to discriminate between iron (II) and (III); the results will be impacted by the way the water is sampled, preserved, and analyzed.

Other than minimizing polymer degradation, it is important to inject consistently at the target viscosity. This can be achieved with periodic sampling or by inline viscosity monitoring. Checking the water salinity periodically is important because any change can greatly impact the viscosity of the polymer solution injected.

Water treatment is not always required; several strategies can be evaluated, depending on the problem to solve or the contaminant to remove.

6.1.2. Oxygen Removal

The presence of oxygen with contaminants in the water will generate free radicals that will break the polymer chains, leading to a decrease in viscosity.

As soon as the oxygen content in water is above 20 ppb, it is highly recommended to remove oxygen and to blanket the dissolution device and maturation tank with nitrogen, even if there are no contaminants in the water (Fe^{2+} or H_2S). Avoiding oxygen ingress will always be beneficial to polymer flooding, will prevent corrosion generated by high-salinity waters, and will minimize any potential bacteriological development in the reservoir.

Depending on the oxygen concentration and water/polymer chemistry, two main options are available to achieve less than 40 ppb in the water and polymer solution:

- *De-aeration.* If tthe oxygen content is above 1 ppm, or if chemicals cannot be used to remove oxygen, mechanical deaeration is recommended. Generally, this will help decrease the oxygen concentration to 100 ppb. The standard recommended process is to flow through a packed column under vacuum with countercurrent nitrogen flow for oxygen stripping.

- *Oxygen scavenger.* The addition of an oxygen scavenger is normally considered when the oxygen level in the water is below 1 ppm. Single injection under turbulent conditions into the water stream is the standard way to treat the oxygen present in the water. Generally, this will help decrease the oxygen concentration to less than 40 ppb.

After oxygen removal, in order to ensure that the O_2 concentration stays below 20 ppb, blanketing the dissolution device and maturation tank with high-purity nitrogen (99.9% purity) is mandatory. If oxygen was removed with scavengers, it is compulsory to make sure no oxygen leaks occur: the reaction between the scavenger and oxygen will degrade the polymer. Additionally, any amount of oxygen introduced by the powder or the emulsion in the solution will be compensated for by injection of a scavenger and the high level of purity of the nitrogen used for blanketing, ensuring a maximum concentration of 20 ppb in the solution.

6.1.3. Requirements for Design

The general workflow leading to field implementation starts with polymer selection. Once a polymer concentration has been agreed upon and the pattern(s) has(have) been selected, a conceptual study can be envisioned, after which a basic design is initiated. Constant discussions between stakeholders help move from the basic design to a detailed design, at the end of which procurement and construction are started.

To begin designing an injectivity test, a pilot injection, or larger facilities, some parameters are required to size the injection units, which can be divided into several categories.

Table 0

Category	Parameters required
Field and reservoir	Onshore/Offshore CEOR type (P, SP, ASP) Total injection flow Injection pressure Number of injection wells, and injection strategy per well Oil characteristics
Chemicals	Polymer form (powder or emulsion) Polymer concentration, viscosity Supply (bags, bulk, etc.) Other chemicals and their characteristics and concentrations
Water specifications	Supply pressure Supply temperature Any pretreatment required? Salinity and hardness Solid content and size distribution Oil-in-water content Oil droplet size (if water treatment is required) Oxygen content Iron content H_2S content
Facilities	Type of installation (skid, containers, on-plot systems, etc.) Available footprint Available power Available air, nitrogen, etc. Area classification (explosion proof) Specifications

Conceptual studies are helpful to frame a global injection project and include the following:

- Process and equipment description.
- Preliminary footprint.
- Estimation of the required power supply and utilities.

- Field injection and deployment philosophy.
- Preliminary capital expense (CAPEX), operating expense (OPEX), and schedule.
- Chemicals delivery, storage, and logistics.

The choice of a deployment strategy is critical to minimize capital expenditures and adapt the injection strategy to local field constrains. Two approaches will be discussed in Section 6.5: a modular deployment approach and large, centralized injection plants.

6.1.4. Powder vs. Emulsion

EOR projects typically require large quantities of chemicals, whatever the method: polymer, surfactant–polymer, or alkali-surfactant-polymer (P, SP, or ASP) (generally above 50–100 kg/h). The use of polymers in inverse emulsion (liquid form) will require standard pumping processes followed by inversion and dilution steps. For polymer powder (solid granular form), pneumatic and/or mechanical processes will be necessary for transportation before the hydration and maturation steps, the latter requiring specific equipment. This difference is critical for offshore operations, where powders are not currently used in large quantities in EOR applications.

The polymer powder active content is usually 90%, compared with the emulsion, which conventionally ranges from 30% to 50%. Given the density of the product (0.7 for the powder and 1.05 for the emulsion), the required storage volume will be twice as large for a 30% active emulsion compared to the powder. Large buffer storage and footprint will be required for the liquid form. One existing solution would be, where applicable, to use the hull tanks of floating production storage offloading (FPSO) vessels. Onsite facilities (excluding storage) will be significantly smaller for emulsion compared with powder.

6.2. Injection Equipment for Emulsions

When using polymer in emulsion form, inline static mixers (with a significant pressure drop) or a mixing pump are required to invert the water-in-oil emulsion (W/O) into an oil-in-water emulsion (O/W), which is then injected into the reservoir. The inversion can be performed directly at the target concentration or in a two-step process:

- Inversion between 10000 and 20000 ppm using a static mixer, by applying a significant pressure drop from 8 to 15 barg.

- Dilution to the target concentration using another static mixer and by applying a more standard pressure drop from 1 to 3 barg (to avoid shear degradation).

6.3. Injection Equipment for Powders

Polymer in powder form can be delivered in bags (25 or 750 kg) or bulk containers. The active matter is usually above 88% and grain size smaller than 1 mm. The first step is dissolution in the injection brine to make a so-called *mother solution*; this is a viscous solution containing up to 2% active polymer. Dispersion and agitation are required to achieve full polymer dissolution and avoid the occurrence of undissolved particles – *fish eyes* – that could impair injectivity (Figure 6.1).

6.3.1. Dispersion and Dissolution

The common dissolution process is generally composed of two steps: powder mixing with water, followed by maturation in dedicated tanks. The mixing phase requires sufficient energy to ensure that polymers grains are properly wetted by water, avoiding the formation of fish eyes that will not dissolve in the long term.

Figure 6.1 **Example of a skid for polymer preparation**

1. Big-bag unloading station

2. 2 m³ polymer powder silo

3. Polymer slicing unit: Floquip PSU 300

4. 20 m³ maturation tank

5. Nitrogen blanketing

In the past, mixing was done by conventional dissolution systems such as eductors. Those systems, widely used in other polymer applications, have several limitations when it comes to EOR:

• Difficulty processing large quantities of polymer at concentrations above 0.5%.

• Long dissolution due to a lack of energy during mixing.

• Oxygen ingress or the need for large nitrogen volumes to minimize it.

• Filtration requirement due to the presence of gels and fish eyes.

The concentration of the mother solution and maturation times will greatly impact the size of the facilities. To overcome these limitations and minimize capital expenditures as well as footprint, SNF has developed and patented a high-energy mixing system called a polymer slicing unit (PSU) [1-4]. This system has the following features (Figure 6.2):

Floquip polymer slicing unit – overview

Figure 6.2

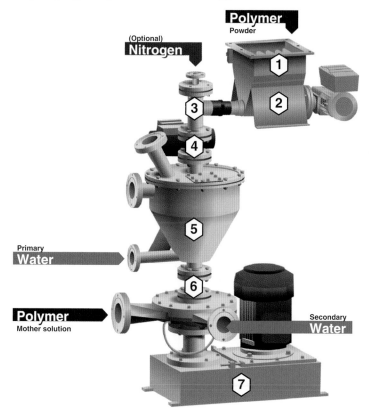

1. Magnetic grid

2. Dosing screw feeder / Gear motor

3. Flexible connection (weighing)

4. Feeder isolation valve

5. Wetting tunnel

6. Cutting head (rotor / stator)

7. Electrical motor / Pulleys / Belt

- High concentration of the mother solution (1% minimum, 1.5% as a standard, and potentially up to 2% in some cases).

- Reduction of the maturation time by a factor of 2 compared to conventional systems, down to 30 minutes.

- No fish eyes generated, so no filtration required of the polymer solution prior to injection.

Therefore, the size of the equipment downstream of this device, such as maturation tanks, pumps, etc., can be tremendously reduced (divided by 4), allowing the deployment of standard skid-mounted units including the entire dissolution and hydration system.

Another important aspect is the possibility of blanketing the entire system with nitrogen to avoid oxygen ingress and therefore limit chemical degradation. Nitrogen can be generated on site with compressors using ambient air, with a purity above 99.9% and limited consumption thanks to the optimization of the overall dead volumes in the installations and pipelines.

6.3.2. Maturation

Maturation of the polymer solution is achieved in an agitated tank with sufficient residence time to allow full dissolution and viscosity development. To minimize footprint, the maturation tank is divided into four agitated compartments. Low-speed agitators are used to avoid mechanical degradation (3 m/s for the maximum tip speed is a rule of thumb). The maturation tank headspace is blanketed with nitrogen to avoid oxygen ingress.

6.4. Field Development Approaches Onshore

For a polymer flood pilot, the dissolution and injection equipment is usually installed at the same location (and generally in the vicinity of the injection wells). This configuration for pilot projects is possible since the footprint required and the number of injection pumps are generally limited.

Additionally, for such projects, placing all the equipment at a single location facilitates operations and monitoring of the injection parameters (Figure 6.3).

Picture of the inside of a skid (PSU at the front, maturation tanks in the back)

Figure 6.3

When it comes to full-field deployment (or partial full-field, i.e. phased development/implementation), the number of injection wells and the size of surface facilities will increase substantially. The most convenient and logical approach would be to build centralized polymer facilities, including polymer storage, dissolution, and maturation. This central polymer facility could be built all at once or progressively by adding modules to increase overall capacity. The location of high-pressure injection equipment could be either close to the central polymer facility or remotely in the field near the injection wells. A modular approach (for polymer preparation and injection) favors progressive investment.

Several injection configurations can be considered, based on existing field facilities, injection philosophy, and operation constraints. The main questions impacting this development are as follows:

- Is there already a waterflood running in the field where polymer flooding will be implemented?

- Is high-pressure water available? Low-pressure water?

- Is the injection strategy one pump per well or one pump for several wells with non-degrading chokes?

- Is the company considering a modular deployment or a centralized facility?

6.4.1. Existing Waterflooding in the Field

When the field is under waterflood, the easiest configuration is to distribute polymer mother solution to each well from a central dissolution facility.

A central dissolution facility comprises at least the following items:

- Polymer powder storage.

- Polymer dissolution and maturation facilities.

- Polymer mother solution metering pumps.

The distribution of the polymer mother solution to the wells can be achieved by two different means, discussed next (the first of which is usually the preferred option). Mother solution high-pressure injection pumps can be located either in the same area as the central dissolution facility or remotely in the field (at the well pads, for example).

6.4.1.1. One Pump per Well – Injecting Mother Solution

The mother solution is pumped at high pressure by a group of reciprocating pumps. Pressurized mother solution is then sent to the different wells through a common header (or more, if required by field configuration), distributed to each of these wells by a dedicated low-shear control device that individually controls flow/pressure of solution per well, and finally injected into the existing waterflooding line (downstream of the water-flow control valve).

This solution is preferred since it limits the number of high-pressure pumps (one pump will feed several wells).

6.4.1.2. One Pump for Several Wells – Injecting Mother Solution

The mother solution is pumped at high pressure to each well by a dedicated reciprocating pump. Pressurized mother solution is then sent to the well through a dedicated flowline. Flow/pressure are controlled by each high-pressure pump.

This solution is not preferred, especially when the number of injection wells is important, because it requires one pump and one polymer flowline per well.

6.4.2. No Existing Waterflooding in the Field

When there is no existing water injection system, the easiest configuration is to distribute the diluted solution (ready to be injected) to each well from a central dissolution facility.

Such a central dissolution facility comprises at least the following items:

- Polymer powder storage.
- Polymer dissolution and maturation facilities.
- Polymer mother solution metering pumps.
- Post-dilution of mother solution.

The distribution of the diluted solution to the wells can be achieved by two different means, discussed next (the first of which is the preferred option). The high-pressure pumps for the diluted solution can be located either in the same area as the central dissolution facility or remotely in the field (at the well pads, for example).

6.4.2.1. One Pump per Well – Injecting Diluted Solution

The mother solution is post-diluted with water at low pressure. The diluted solution is pumped at high pressure by a group of reciprocating pumps. The pressurized diluted solution is then sent to the different wells through a common header (or more, if required by field configuration), and distributed to each of these wells by a dedicated low-shear control device that will individually control injection flow/pressure of solution per well.

This solution is preferred because it limits the number of high-pressure pumps (one pump will feed several wells).

6.4.2.2. One Pump for Several Wells – Injecting Diluted Solution

The mother solution is post-diluted with water at low pressure. The diluted solution is pumped individually at high pressure for each well by a dedicated reciprocating pump. Pressurized diluted solution is then sent to the well through a dedicated flow-line. Flow/pressure are controlled by each high-pressure pump.

This solution is not preferred, especially when the number of injection wells is important, because it requires one pump and one polymer flowline per well.

6.4.3. Logistics for Onshore Projects

Onshore, polymer in powder form can be delivered in 25 kg bags, 750 kg big bags, or in bulk, depending on the size of the project (Figure 6.4).

Logistics for powder

Figure 6.4

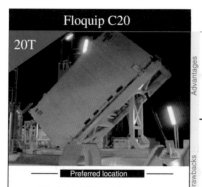

Floquip C20

20T

Advantages
- Closed-loop system (no moisture ingress)

Drawbacks
- Tilting mechanism required
- Limited by the capacity of the CEOR vessel crane
- Crane operations

Preferred location
- Designed for onshore
- Adaptable to offshore

Bag-in-box

20T

Advantages
- Transportation cost

Drawbacks
- Major HSE risks while unloading
- Tilting mechanism required
- Waste bags
- Risk of bag laceration

Preferred location
- Onshore

Floquip OT

16T

Advantages
- Closed-loop (no moisture ingress)
- No need for a tilting mechanism

Drawbacks
- Cost of the unit
- Crane operations
- Limited by the capacity of the CEOR vessel crane

Preferred location
- Offshore

PD truck

45T

Advantages
- Closed-loop system (no moisture ingress)
- High unloading flowrate

Drawbacks
- Only land transportation
- Cost of transportation

—— Preferred location ——
- Onshore

Railroad car

90T

Advantages
- Closed-loop system (no moisture ingress)
- High unloading flowrate
- Large capacity

Drawbacks
- Railway transportation only (requires unloading hub)

—— Preferred location ——
- Onshore

Supply boat

100 to 1000 T

Advantages
- Limited operations onto the CEOR vessel
- High unloading flowrate

Drawbacks
- O&M management

—— Preferred location ——
- Offshore

6.5. Key Considerations for Offshore Implementation

The constraints are very different when designing an onshore or offshore polymer flooding project. The keys for offshore design are the space and load limits of the FPSO or platform. Local weather conditions and logistics will also dictate the form of the product to be used (powder, emulsion liquid, etc.). Polymer in inverse emulsion form is much easier to handle offshore, particularly when weather conditions are unfavorable. It does not require specific dissolution systems: the inversion is performed on the fly with sufficient energy, and the solution is then diluted and injected into the reservoir. Polymer in powder form has a higher active content (up to 90% compared to the 30 or 50% for the emulsion), is cheaper to implement, and requires less logistics.

The main points to consider for offshore applications are the following:

- Specific design considering footprint limitations and load constraints.

- Modular solution to limit installation activities on the FPSO, platform, or any vessel. These modules must comply with lifting/handling constraints such as crane limitations and specific offshore procedures.

- Sensitivity to vibrations during operations (pumps running, conveying, etc.). Specific attention must be paid to load interactions and structural issues.

- Hazardous classified area. Equipment and modules must be adapted to hazardous classification constraints, especially for the control room, motor control center, and utilities shelters.

- Environmental conditions, including the effect of vessel motions (rolling, sagging, pitching, heaving), wind loads, and blast loads on structural and equipment design.

- Corrosion. Protection against the corrosive atmosphere (stainless steel equipment, specific painting procedures, nitrogen blanketing, and greasing).

- Waste management and back-produced water. Use of chemicals offshore and possible waste to be handled. Specific cleaning procedures should be studied.

- For offshore projects, non-shearing injection pumps will transfer the injected solution through the main risers up to the several subsea Christmas trees dedicated to each well. Choke valves will then be used to control the flow and the pressure for each well. Specific solutions should be selected to minimize viscosity degradation.

- Logistics. For chemical EOR, a significant quantity of chemicals must be handled in either liquid or powder form. Therefore, the supply chain has to be secured accordingly.

Table 6.1 shows basic calculations for a standard pilot with the injection of 30 000 bpd of polymer solution at 1500 ppm and a larger, full-field injection at 200 000 bpd.

Table 6.1 Comparison between powder and emulsion quantities required for two field cases

Project	Pilot		Full field	
Injection rate (bpd)	30 000		200 000	
Chemical form	Powder	Emulsion	Powder	Emulsion
Concentration (ppm, active)	1 500		1 500	
Daily consumption (mT, active)	7.2		47.7	
Daily consumption (m³/d, commercial)	10.3	22.9	68.1	151.4
7 d buffer storage (m³/w)	72.1	160.3	476.7	1059.8

The essential learning is that logistics for these offshore CEOR projects is one of the most critical parameters, in addition to the footprint requirements. The delivery of the polymer to the field storage silos/tanks can be carried out by various means, depending on the following:

- Daily consumption (from a few tons per day to several hundred).

- Duration of operations (from a few weeks to 10 or 20 years).

Logistics for EM

Figure 6.5

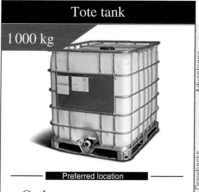

Tote tank
1 000 kg

Preferred location
- Onshore
- Offshore (short-term duration)

Advantages
- Transportation cost
- Flexibility

Drawbacks
- Forklift and manual operations
- More appropriate to onshore application

Transit tank
1–10 T

Preferred location
- Onshore

Advantages
- Fitted to offshore crane weight limitations

Drawbacks
- Crane operations
- Offshore connection operations

- Polymer form: powder or emulsion (decided based on field, budget, and operational constraints).

- Oil field location: onshore/offshore, distance from the main polymer storage warehouse, existing infrastructure, etc.

- Delivery frequency, size of the final storage silos/tanks, and accessibility.

Based on the project requirements and constraints related to the polymer form, some specific solutions exist (Figure 6.5).

For all of the offshore and part of the onshore projects, a warehouse with a few weeks of polymer storage is required.

Isotank	Advantages	• Closed-loop system • Large capacity
25 T	Drawbacks	• Limited by the capacity of the offshore crane • Crane operations for offshore
Preferred location • Onshore • Offshore		

Supply boat	Advantages	• Limited operations offshore • High unloading flowrate
100–1000 T	Drawbacks	• O&M management • Costs
Preferred location • Offshore		

This will allow the polymer manufacturer to deliver the product using the least expensive logistic solution:

- Big bags stacked in 40' containers for the powder form.

- 25 mT isotanks for the emulsion form.

- Other specific bulk options that can be studied case by case, depending on the site location and volumes required.

6.6. ASP Process

6.6.1. ASP Reminder

Alkali surfactant polymer (ASP) injection aims to mobilize trapped oil in oil-bearing formations by playing both on the interfacial tension and the mobility ratio. The design of a proper formulation is complex and takes time. Moreover, injection water softening is required to remove any divalent cation that could precipitate with the alkali.

6.6.2. Water Softening

One of the major issues to address when designing an ASP flood is the formation of scales. The increase in pH due to the injection of alkali will lead to the precipitation of divalent ions present in the water (such as Ca^{2+} and Mg^{2+}) with the carbonate or hydroxide of the alkali. To prevent or at least significantly reduce scale formation, it is necessary to soften the water used for the ASP process and reduce the concentration of divalent cations, targeting a total hardness less than 5 ppm (sometimes less than 1 ppm).

The systems most commonly used to soften the water in ASP flooding are ion-exchange resins. Two main types of resins can be used:

- Strong acid cation (SAC)

- Weak acid cation (WAC)

The exchange process trades monovalent hydrogen or sodium ions on the resin matrix for calcium and magnesium ions in the water. Several factors influence the choice between WAC and SAC exchange for softening water:

- Total hardness
- Total dissolved solids
- Cost of regeneration chemicals
- Capital investment
- O/W content

WAC resins can operate at total dissolved salts (TDS) up to 30 000 and 500 ppm hardness, while SAC resins are normally acceptable for water with a TDS of less than 5000 ppm. At a salinity approaching seawater or above, SAC resin systems are compromised because of the high sodium content in the water. This type of resin uses concentrated sodium chloride brine for regeneration and has a strong economic advantage in terms of OPEX versus the WAC resin. The latter consumes hydrochloric acid and sodium hydroxide for regeneration, which are more expensive than salt. The cost for regeneration is typically 4 to 10 times greater for WAC resins compared to SAC resins. CAPEX and equipment costs are also two to three times more for WAC compared to SAC. Obviously, the greater the hardness of the water, the greater the impact on OPEX.

The alternatives to resins are prefiltration and reverse osmosis. Reverse osmosis is TDS sensitive and removes monovalent ions. Salt can be added afterward if required. Microfiltration requires resin polishing but does not remove monovalent ions.

If the water contains oil carry-overs, even at a low concentration, resin fouling will likely occur for both WAC and SAC type resins. In that case, only WAC is suitable, since the sodium hydroxide used during the regeneration step will help clean the resin and delay complete fouling.

6.6.3. Chemicals

6.6.3.1. Alkali

In most cases, the alkali used in ASP processes is either sodium carbonate (soda ash) or sodium hydroxide (caustic soda).

6.6.3.1.1. Caustic Soda

Caustic soda is generally sold and delivered as a 50% active liquid. At this concentration, a minimum temperature of 18 °C should be maintained for storage. If this is not possible, the product should be diluted to 20%. Pipe and tank materials should be adapted to use this base. Its use requires a simple water dilution line in which caustic is added to the desired concentration.

6.6.3.1.2. Soda Ash

This product is a little more complicated to handle, because it is delivered in powder form – requiring dissolution – and because water temperature will dictate the maximum concentration achievable and the size of the equipment. Above 35 °C, a 32% saturated solution is possible. Below that temperature, the crystals formed will be difficult to re-dissolve. Different options are possible for the dissolution phase:

- *Continuous mixing process.* Soda ash in powder form is stored in a specific silo or hopper (depending on requirements). From this storage, the powder is metered by a screw feeder into a mixing tank with water. This mixing tank is generally composed of two compartments. The first compartment is used for dissolution of the soda ash in water and is sized to achieve 15 minutes' residence time and ensure proper dissolution. The second compartment acts as a buffer tank to feed the pumps that will meter the soda ash solution into the process. Transfer from the dissolution compartment to the buffer compartment is achieved by overflow. With this process, concentration of the soda ash solution should be selected to remain below the saturation value.

- *Slurry*. The benefit of this process is that storage and mixing are combined in a single tank. Soda ash in powder form is directly unloaded from its packaging (big bags, bulk truck, etc.) into the slurry tank. For large quantities, a buffer silo can be used. The make-up water is introduced into the tank at the bottom through a sparger network that distributes the fluid evenly across the tank. As the dilute liquid migrates up through the crystal bed, it begins to dissolve soda ash crystals. Depending on the demand, the injected water must become saturated (32% active) by the time it reaches the top of the crystal bed. If the liquid fails to reach saturation, a recycle loop can be used to complete the process. The saturated soda ash solution is drawn off behind a baffle that prevents crystals from reaching the suction pipe during the unloading process. The solution used to create the suction at the unloading eductor is also drawn off behind the quiescent zone baffle. The temperature of the slurry storage system is a key parameter in selecting a slurry technology: it should always be maintained above 35.4 °C to prevent formation of hepta and deca hydrates.

6.6.3.1.3. Filtration

Usually, soda ash contains a certain percentage of insoluble particles that must be removed prior to the injection process. In most cases, this is done by inline filtration of the soda ash solution with bag filters. Depending on the purity of the soda ash powder, several steps may be required. Usually, two-step filtration (100 µm and then 5–25 µm) is sufficient.

6.6.3.2. Surfactant

The surfactants that can be used in EOR processes display a wide range of properties. Although some are delivered in liquid form, it is not unusual to encounter mixtures having the consistency of a paste. The key factors regarding surfactant handling are pour point and viscosity; viscosity is sometimes very high, requiring dilution to maintain pumpability. Additionally, heating of storage tanks may be required to reduce apparent viscosity and facilitate pumping.

In some cases, because some surfactants tend to segregate, the product must be re-homogenized in the tank by recirculation or using agitators. Surfactant dilution is a simple process requiring a metering pump to adjust the concentration in the water stream. A static mixer or, in some specific cases, a dynamic mixer, can be added to enhance the homogeneity of the cocktail (the guidelines are given by the manufacturer) (Figure 6.6).

View of a surfactant-polymer injection pilot site

Figure 6.6

Photo courtesy of MOL

6.6.4. Mixing of All Products

Generally, mixing of all ASP components (polymer, alkali, and surfactant) is done through slip-stream injections into the main water stream and supported by static mixers. The order of injection of the different chemicals is dictated by the ASP cocktail designer.

6.7. From the Dissolution Point to the Wellhead

Monitoring the quality and viscosity of the injected solution can be performed with several specific tools [8]. It is also important to design the facilities so that degradation is minimized throughout the process.

6.7.1. Viscosity Monitoring

The main difficulty with manual sampling lies in the procedure used to minimize chemical or mechanical degradation, which can occur during both sampling and the measurement itself. A specific sampling device composed of two cylinders with stabilizing agents has been developed to ensure that no degradation occurs during the measurement (otherwise, samples must be analyzed in a glove box). However, this method requires manpower and can be difficult to operate offshore because waste is generated.

A device developed by SNF measures viscosity inline at low shear rate, high pressure (up to 200 barg), and high temperature (up to 80 °C); the result can be extrapolated to a yield viscosity [1, 5, 9, 10].

The device combines the following:

- A micro-gear pump that can withstand high pressure (up to 400 barg) at its suction and discharge. A variable-speed drive controls the flow rate of this pump.
- A calibrated tube that creates a pressure drop.
- A Coriolis flowmeter.
- A pulsation dampener or absorber.

The fluid is directly and continuously sampled via a bypass in the main pipe under high pressure. The solution is filtered with 25-µm cartridge filters to protect the internal mechanism of the micro-gear pump. The differential pressure transmitter measures the pressure drop across the calibrated tube. The solution is then reinjected in the main pipe. The pressure drop measured under high pressure is extrapolated by the control system to the corresponding viscosity of the water-soluble polymer solution, measured at atmospheric pressure with a laboratory viscometer in the same conditions of concentration and salinity (Figure 6.7).

Inline viscometer

Figure 6.7

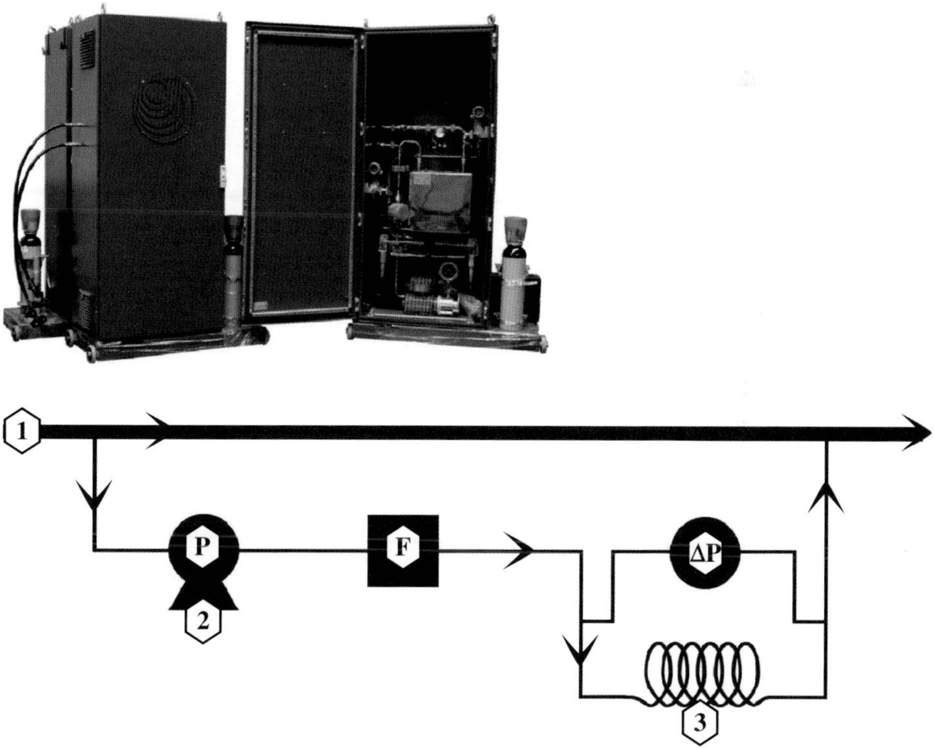

F. Flow meter	1. Main pipe
P. Pump	2. Volumetric pump 20 l/h
ΔP. Pressure drop	3. 20 m calibrated coil

6.7.2. Non-shearing Chokes

In classical waterflooding systems, injection flow/pressure for each well is controlled by an individual device called a *choke valve*. In EOR processes where polymer solutions are used, using a standard choke valve is not recommended, due to the mechanical degradation induced by the high shear generated through the choke as soon as the pressure drop is higher than 5 bar.

SNF has designed non-shearing chokes such as Floquip VPR to provide stable, accurate control of the polymer injection without significant degradation of the solution [6, 7]. The equipment can be either a fully manual system with manually operated valves or a fully automatic system with air/electrical-operated control valves.

These units are designed on a case-by-case basis, using actual injection parameters (polymer type and concentration, temperature, water composition, flow and pressure ranges, etc.).

The unit is composed of two main parts:

- A series of restriction orifices (generally 5–7).

- A series of valves (needle or gate type).

In a Nutshell

Different injection strategies are possible to minimize capital expenditures for the development of CEOR processes. A modular approach, for instance, will allow progressive investment, development, and implementation. A very important aspect to take into consideration is logistics: for full field development, this can be a limiting factor.

References

[1] Pich, R. and Joeronimo, P. (2008). Installation for enhanced oil recovery using water-soluble polymers, method implementing same. Patent WO/2008/071808, filed 25 March 2008 and issued 19 June 2008.

[2] Pich, R. and Joeronimo, P. (2008). Device for preparing a dispersion of water-soluble polymers and method implementing the device. Patent WO/2008/107492, filed 25 March 2008 and issued 12 September 2008.

[3] Joeronimo, P. and Pich, E. (2011). Improved device for dispersing a water-soluble polymer. Patent WO/2011/107683, filed 8 February 2011 and issued 9 September 2011.

[4] Bonnier, J. and Pich, E. (2016). Improved device for dispersing a water-soluble polymer. Patent WO/2016/156320, filed 29 March 2016 and issued 6 October 2016.

[5] Favero, C. and Rivas, C. (2016). Device for in-line monitoring of the quality of a water-soluble polymer solution manufactured from invert emulsion or powder of said polymer. Patent WO/2016/142623, filed 9 March 2016 and issued 15 September 2016.

[6] Remy, P. and Pich, E. (2015). Apparatus for controlling injection pressure in offshore enhanced oil recovery. US Patent 9328589B2, filed 12 February 2015 and issued 3 May 2016.

[7] Soucy, B. (2012). Linear pressure reducer for regulating injection pressure in an enhanced oil recovery system. US Patent 8607869B2. filed 22 November 2012 and issued 17 December 2013.

[8] Pich, R. and Jeronimo, P. (2011). Method of continuous dissolution of polyacrylamide emulsions for enhanced oil recovery (EOR). US Patent US8383560B2, filed 19 May 2011 and issued 26 February 2013.

[9] Bonnier, J., Rivas, C., Gathier, F., et al. 2013. Inline viscosity monitoring of polymer solutions injected in chemical enhanced oil recovery processes. Paper SPE 165249 presented at the SPE Enhanced Oil Recovery Conference, Kuala Lumpur, Malaysia, 2–4 July.

[10] Quillien, B. and Pich, E. (2012). Device for measuring and controlling on-line viscosity at high pressure. Patent WO/2012/140092, filed 11 April 2012 and issued 18 October 2012.

Chapter seven

Produced Water Treatment

The treatment of effluents of oil and gas operations can sometimes be impacted by the presence of produced polymer. The current technologies and developments will be highlighted in this chapter.

Essentials of Polymer Flooding Technique, First Edition. Antoine Thomas.
© 2019 John Wiley & Sons Ltd. Published 2019 by John Wiley & Sons Ltd.

7.1. Introduction

In hydrocarbon-bearing formations, water is very often associated with oil and/or gas. After a period of time in production, this water – called *formation water* – will likely be produced along with the oil and/or gas at an increasing percentage (called the *water-cut*) until little hydrocarbon is produced. In order to maintain the pressure in the reservoir and displace the fluids of interest, a common practice is to inject water into the formation over an extended period until the process is rendered uneconomical by the quasi-absence of oil production. The result is the production of large quantities of water that require handling, treating, disposal, or reinjection. The associated treating and handling costs can be significant, ranging from a few cents to a few dollars per barrel, depending on the treatment facilities, oil density/viscosity, and other contaminants such as solids, forcing companies to develop strategies to tackle this issue.

Polymer flooding is one of the techniques that helps delay water breakthrough when implemented early enough. Even when this technique is deployed late in the life of the field, it can positively impact fluid ratios at the production side by decreasing the water-cut and increasing the oil cut, resulting in substantial savings on water handling.

A question that often arises is the possible impact of polymers on surface facilities during oil/water separation and, afterward, on the water treatment facilities. This chapter will discuss these aspects, and possible solutions will be outlined. However, case-by-case approaches remain necessary to find the most economical solutions.

The most difficult aspect of designing appropriate facilities in the case of polymer injection is to predict the polymer's characteristics (hydrolysis, molecular weight) and breakthrough in the producers. This depends on several design parameters:

- How much pore volume is injected? If less than one reservoir pore volume is expected to be filled by the polymer slug, then it is very unlikely that the original polymer concentration will ever be reproduced. The same is true for viscosity.

- What will happen when switching back to water? Again, if less than one reservoir pore volume is expected to be filled by the polymer slug, then, when switching to water, the water probably will not push the polymer slug evenly, but rather will finger through it and lead to dilution. Also, the polymer will progressively be stripped from the chase water by adsorption in the virgin zones of the reservoir. In the producers, the concentration will therefore increase slowly, peak, and then decrease with a very long tail.

- What is the viscosity injected? A high polymer viscosity will greatly improve sweep efficiency, enlarging areal and volumetric sweep efficiencies and therefore delaying breakthrough. Retention is also satisfied more quickly when a higher concentration and viscosity is injected.

- What will be the characteristics of the produced polymer? How will the chemistry and molecular weight be changed during transit in the reservoir and through the production facilities? Researchers have observed that, in some cases, polymers experience changes in hydrolysis degree (increase) and molecular weight (decrease). This aspect will greatly impact the viscosifying and flocculating properties of the polyacrylamide.

Viscosity in the producer is very difficult to predict since the polymer can be rapidly degraded by the lifting systems or pumps, notwithstanding dilution within the reservoir. It is slightly less complicated to predict polymer concentration, especially with a proper 3D model and adequate keywords. As a rule of thumb, we can say that the maximum polymer concentration in the producer will be a function of the following dimensional parameters:

$$C_{polymax} = C_{poly\,inj} * PV_{inject} * \textbf{retention factor} * \textbf{reservoir swept}$$

With $C_{polymax}$ = maximum polymer concentration likely to be observed in the producers; $C_{polyinj}$ = injected polymer concentration; retention factor = polymer retention (can be defined as a percentage of polymer lost); and reservoir swept = percentage of reservoir containing the polymer contacted by the water chase.

All parameters except the initial concentration injected have values between 0 (excluded) and 1. Values of theoretical produced polymer concentrations should therefore be much lower than the initial concentration injected, but the actual concentration is subject to flow anomalies such as channels or fractures that would otherwise contribute to earlier breakthrough of higher produced polymer concentrations. Locations where this may be problematic should be identified through tracer testing. In general, the breakthrough time can be approximated by looking at the production history, especially the time it took for the injection water to break through during the secondary waterflood.

An important aspect to consider is possible dilution when all the effluents from different production areas are pumped to the same central processing facility, especially when not all of these zones are under polymer injection. In that case, dilution will help drastically decrease the viscosity and concentration in the produced fluids, limiting the impact on the existing facilities.

7.2. Generalities

7.2.1. Produced Water Characteristics

The produced water composition will vary over the life of the field due to factors and interventions such as water injection (with composition different from formation water), reservoir stimulation, introduction of chemicals, bacterial activity, enhanced oil recovery (EOR) techniques, and so on.

Produced water is basically a mixture of water and hydro-carbons and can also contain a variable percentage of the following:

- Suspended solids, including clays
- Production chemicals
- Scales
- Corrosion products
- Traces of heavy metals
- Dissolved organics (hydrocarbon included)
- Dissolved gas (H_2S)
- Dissolved iron
- EOR chemicals

Proper treatment is required before reusing or disposing of the produced EOR fluid. Various technologies exist, some of which will be described in the upcoming sections.

A polymer can have several impacts on the water treatment facility if/when it breaks through:

- Polymers act as a flocculant and can agglomerate suspended solids.

- A polymer can increase the viscosity of the water phase – assuming it has the right molecular weight and chemistry – therefore impacting the treatment facilities, where separation mechanisms are governed by gravity and whose performance can be understood and described using Stokes' law.

- A polymer acts as a friction reducer, decreasing drag at Reynolds numbers above 3000. It can partly suppress vortices in specific equipment, such as hydrocyclones.

- A polymer is an ionic macromolecule that can interact with charged particles or electrolytes, sometimes enhancing droplets' coalescence.

As a side note, polyacrylamides are strictly water-soluble polymers. Unpublished studies showed that this polymer family was not present in crude oil and did not impact the crude refining processes. The possible influence on water treatment facilities will be discussed in the next sections.

7.2.2. Oil and Gas Processing

The principal function of surface facilities is (i) to separate oil, gas, water, and solids to deliver hydrocarbons meeting sales specifications and (ii) to dispose of the water.

The first step in this process is separating water from oil and, if present, gas from liquid. This is conventionally done in separators with various designs, using gravity, and heat where necessary, as the main drivers. Oil and gas are directed to specific treating facilities to polish oil to sales specifications, while water enters a secondary process in which the remaining oil and solids are removed before disposal.

This secondary process can involve various technologies based on filtration, gravity, coalescence, etc. A brief description will be given later of the possible impacts of polymers on overall efficiency.

7.3. Oil and Gas Separation

Two technologies – whose efficiency can be impacted by polymers – will be specifically discussed here.

7.3.1. Separators

Separators can be horizontally or vertically orientated vessels and are usually composed of several devices or separation aids such as cyclones, filters, and plate packs (Figure 7.1). The separation performance depends on many factors such as flow rates and fluid properties. The main driver for separation is

Free-water knockout tank

Figure 7.1

gravity: inside the *gravity/coalescing zone* (often the free-water knockout tank), the oil droplets separate from the water thanks to density differences, as expressed by Stokes' law:

$$V_r = \frac{(\rho w - \rho o) * g * D2}{18\mu_w}$$

where V_r = drop/rise velocity (m/s); ρ_w = water density (kg/m³); ρ_o = oil density (kg/m³); μ_w = water dynamic viscosity (kg/(m s)); g = gravitational acceleration (9.81 m s⁻²); and D = drop diameter (cm).

Analysis of Stokes' law reveals that polymers can impact both viscosity (of the water phase) and droplet size by affecting coalescence. Predicting the viscosity of produced water containing polymers is very difficult, especially if the production equipment degrades the polymer backbones via mechanical shear, as is the case with some pumps and valves.

In theory, there are two ways to improve separation in cases where significant viscosity is produced:

- By breaking the viscosity via shear (or other methods). A drawback can be the creation of more stable emulsions.

- By increasing residence time in the separator. However, equations show that doubling the residence time will cause an increase in droplet diameter of only 19% [1].

Internal studies and field tests have shown that, generally speaking, polymers do not participate in oil emulsification and do not create intermediate rag layers. Also, the proportions of the phases obtained and the separation time are not affected by the presence of polymers. However, the compositions of the water-dominant phase and oil-dominant phase could be impacted either positively (better crude dehydration) or negatively (worse crude dehydration and higher oil content in the water) by the presence of polymers.

With regard to droplet size, studies performed with crude from the Daqing fields [2, 3] showed that polymers could enhance the coalescence of droplets below a certain concentration (and therefore a certain viscosity).

7.3.2. Heater Treaters

In a typical separation process, the produced fluid goes through three chambers: a high-pressure separator, a low-pressure separator, and a dehydrator. In the dehydrator, the wet oil is heated to reduce the viscosity of the continuous oil phase. This equipment can be categorized as mechanical or electrostatic dehydrators [4]. A mechanical heater is composed of a fire tube followed by a mechanical gravity separator. Electrostatic dehydrators use electrostatic fields to mobilize dispersed water without mobilizing the continuous oil; this is based on the fact that oil is almost nonpolar and nonconductive, whereas water is a polar and conductive molecule.

Issues have been reported in heavy oilfields with mechanical heater treaters and failures of fire tubes: a continuous temperature increase of the tube skin was observed after solids deposition occurred. The origin of this skin is probably the polymer flocculating solids in suspension, favoring wall deposition and skin formation. Remediation methods – such as scale inhibitor development, heat flux optimization, and installation of spray bars – were used to limit maintenance time and decrease the number of interventions [5, 6]. Also, a decrease in skin temperature is possible, to maintain operational efficiency [7]. Studies have shown that the use of electrostatic dehydrators is a preferred option in projects with polymer injection.

7.4. Water Treatment

7.4.1. Introduction and Generalities

Once the hydrocarbons have been separated from the bulk of the produced fluid, the water is directed to treating facilities designed to remove the remaining oil and solids. Equipment to remove oil and solids relies on different principles:

- Gravity separation (often with coalescing aids)
- Gas flotation
- Cyclonic separation
- Filtration
- Centrifuge separation

Chemical or biological methods can be introduced into the mix to remove specific components.

7.4.2. Gravity Separation

Gravity-separation devices include American Petroleum Institute (API) separators, skim tanks, skim piles, and plate coalescers. Such equipment is usually rather inexpensive, but it requires a large footprint because of the residence time needed to favor coalescence of oil droplets. Oil concentration and particle-size distribution are needed for the proper design of a gravity separator, which can render the process more complicated when an EOR method such as surfactant–polymer (SP) or alkali-surfactant-polymer (ASP) is implemented.

As discussed in the previous section, the principle of gravity separation can be described using Stokes' law:

$$Vr = \frac{\left(\rho_w - \rho_o\right) * g * D^\dagger}{18\mu_w}$$

where V_r = drop/rise velocity (m/s); ρ_w = water density (kg/m³); ρ_o = oil density (kg/m³); μ_w = water dynamic viscosity (kg/(m s)); g = gravitational acceleration (9.81 m/s²); and D = drop diameter (cm).

Several conclusions can be drawn from the analysis of this equation [1]:

- The greater the difference in density between oil and water, i.e. the lighter the oil, the greater the rising velocity of the oil phase.

- Increasing the value of gravitational acceleration – for example, by centrifugal motion – will increase the separation velocity.

- The larger the oil droplet, the greater its rising velocity to the surface.

- The lower the water viscosity, the easier it is to treat the produced fluid. Temperature is therefore an important factor in water treatment, since it impacts viscosity.

Polymers can have an impact on water viscosity (depending on their concentration and molecular weight) and also on coalescence. Below a certain concentration, polymers can help coalescence by bringing together oil droplets [8]. However, when the concentration and viscosity increase, it becomes more difficult for the droplets to move and contact each other, decreasing both coalescence efficiency and rising velocity.

Water clarifiers (deoilers) are often used to favor the coalescence of oil droplets and facilitate separation. It is paramount to check the compatibility of these chemicals or any coalescing aid with produced polymers, since the polymers can interact with the clarifier and impact the overall efficiency of the process. Bottle tests can be performed to optimize the choice of the deoiler chemistry to maintain good separation efficiency.

7.4.2.1. Deoilers

Deoilers, reverse demulsifiers, coagulants, and flocculants are used to treat produced water containing oil, to enhance droplets' coalescence and accelerate the clarification process (Figure 7.2). Oil droplets usually bear a negative charge at their interface with water; they can be destabilized by adding chemicals with a cationic charge (coagulant) and then flocculated with either an anionic or a cationic product. The presence of residual anionic polymers coming from EOR will impact the dosage of cationic deoiler (linked to the cationic demand of water) needed to properly treat the residual oil [6, 9] (Table 7.1).

Deoiling test – use of optimized chemical compositions

Figure 7.2

Improved deoiling efficiency

Table 7.1 Cationic demand and coagulant addition required as a function of polymer concentration in the effluent

Parameter	Values					
Polymer content (ppm)	0	10	50	200	500	1000
Cationic demand (meq.L^{-1})	10	60–64	230–250	860–900	2090–2240	4120–4410
Coagulant addition (ppm)	10	10	30–32	128–133	279–296	471–494

The main issue with this approach when a polymer is present in the produced water is the quantity of sludge created by the reaction of both chemicals. It has been evaluated that treating a produced fluid containing 800 ppm of produced polymers with conventional deoilers creates 12 kg of sludge per cubic meter of produced water. This strategy should therefore be avoided to minimize the volume of sludge to be treated, handled, and disposed of.

Promising new chemical formulations and strategies have been developed and are being tested to maintain acceptable deoiling efficiencies in the presence of produced polymers [10–13]. These products are mainly nonionic.

7.4.3. Gas Flotation

With dispersed (induced) or dissolved gas flotation units (induced gas flotation [IGF] and dissolved gas flotation [DGF], respectively), large quantities of gas bubbles are injected into the water stream, where they attach to oil droplets, causing them to rise more efficiently to the surface where the oil can be

Induced gas flotation device

Figure 7.3

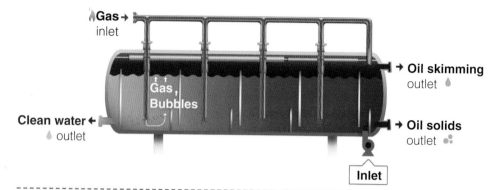

recovered (Figure 7.3). Through gas introduction, oil droplets tend to rise to the surface as a result of three principles [14]:

- Increased vertical velocity of lighter components.

- Breakage of large vortices and conversion to smaller ones. The flow becomes more heterogeneous.

- Attachment of gas bubbles to contaminants, creating a stronger, upward buoyancy force.

An efficient process therefore requires a uniformly distributed flow of small gas bubbles with a velocity low enough to enable attachment to the oil droplets.

Internal studies have shown that IGF and DGF lose appreciable efficiency in the presence of polymers (up to 60%). Several mechanisms might explain the poor performance of the technology in that case:

- In the presence of polymers and possible interactions with droplets, it might be difficult for gas bubbles to attach to oil droplets and make them rise.

- Polymers might not favor droplet coalescence and growth.

- If a certain viscosity remains, where residual polyacrylamides show non-Newtonian behavior, gas channeling can occur. In this case, preferential paths will be created through the fluid, leaving a large percentage of the oil droplets untouched.

- In theory, the size of IGF equipment should be increased proportionally to the increase of effluent viscosity.

Therefore, neither IGF nor DGF should be considered as a standalone treatment when polymers are expected or present in the produced water. A strategy to degrade the produced polymers might also be considered, to minimize side effects of increased produced-water viscosity.

7.4.4. Cyclonic Separation

Hydrocyclones are widely used to deoil produced water: they are insensitive to motion or orientation and have the smallest footprint. Deoilers are pressure-driven: they use fluid-pressure energy to create rotational fluid motion. The main issue with polymers lies in the drag-reducing properties they impart on the water phase: they are known to decrease the pressure drop by homogenizing the distribution of streamlines and smoothing the transition between laminar and turbulent flows. Therefore, it will be much harder to create a pressure drop and vortices when a polymer is present in the water, greatly affecting the process on which hydrocyclones are based. Studies have shown that, in the presence of polymers, the deoiling efficiency of hydrocyclones can be decreased by 60% [1]. The possible stabilization of oil droplets by polymers further erodes deoiling efficiency. Hydrocyclones with gas injection have shown some improvement but have are not seen widespread application in the oil and gas industry.

7.4.5. Centrifuges

Disk stack centrifuges (DSCs) are also used to remove oil in fluids containing up to 1000 ppm of oil-in-water (O/W) at the inlet. Deoiling efficiency up to 85% has been observed with

500 ppm polymers present. However, frequent maintenance is required due to the significant risk of failure caused by mechanical erosion.

7.4.6. Filtration

Different filters are used in the oil and gas industry to remove both solids and oil carry-over. They can be used either to remove the coarse part or for polishing. For the sake of simplicity, we will define two major categories:

- Media filters: walnut-shell filters and sand filters
- Membranes

7.4.6.1. Media Filters

7.4.6.1.1. Walnut-Shell Filters

These filters are designed specifically to remove residual, dispersed hydrocarbons from produced water (Figure 7.4). They are often used as polishing equipment after IGF. No studies have been published on the impact of polymers on deoiling efficiency. However, internal studies have shown promising results; investigations are ongoing to test the limits of this system in the presence of polymers.

7.4.6.1.2. Sand Filters

In addition to the media type, sand filters differ from walnut-shell filters in the method of backwashing. Unpublished studies have shown that this type of filter becomes inefficient when polymers are present in the water.

Several technology providers have worked on new materials and designs to improve deoiling efficiency in the presence of polymers:

- Siemens PerforMedia
- Schlumberger MYCELX technology

Figure 7.4 **Principle of walnut-shell filters**

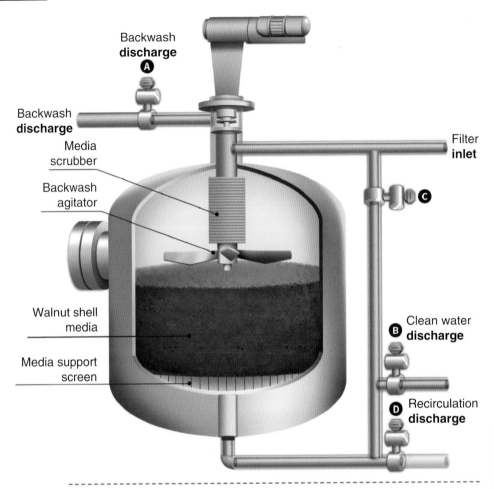

Backwash
discharge
Ⓐ

Backwash
discharge

Media
scrubber

Backwash
agitator

Walnut shell
media

Media support
screen

Filter
inlet

Ⓒ

Ⓑ Clean water
discharge

Ⓓ Recirculation
discharge

7.4.6.1.3. Siemens PerforMedia

This Siemens proprietary synthetic media is designed to handle up to 500 ppm O/W, targeting less than 10 ppm O/W residual in the effluents and removing 90% of solids with a size >10 μm. Yard tests were performed in 2017, in collaboration with SNF, utilizing synthetic effluents and polymers with characteristics close to what could be expected in produced water.

Test conditions	Values	
Flux (m³/hr/m²)	24.4	30.5
Feed O/W (ppm)	241	241
Feed polymer (ppm)	493	484
Effluent O/W (ppm)	6.2	16.5
Oil removal efficiency (%)	97.4%	93.2%

Table 0

In addition to efficiently removing more than 90% of O/W, no change in polymer characteristics was observed during the treatment, opening the possibility to reuse the polymer for making-up and hydrating a new viscous solution for injection.

7.4.6.1.4. Schlumberger MYCELX

The equipment as tested in the yard was composed of a coalescer, a proprietary filter media (Regen), and possibly a polisher. Tests were performed with O/W ranging from 500 to 1300 ppm and polymers with solution viscosities from 3 to 18 cP. Oil-removal efficiency ranged for all tests between 95% and 97.5%, leaving the polymer's characteristics unchanged. Field tests corroborated the results observed in the yard.

7.4.6.2. Membranes

Membranes are usually divided in two categories (Figure 7.5): those that primarily retain particles (micro- and ultrafiltration) and those that primarily retain molecules and ions (nanofiltration and reverse osmosis) [15]. The way water moves through these membranes is different in the two categories. For micro/ultrafiltration (MF/UF), water moves through pores; whereas for nanofiltration and reverse osmosis, water moves through molecular structures. The pore size for MF ranges from 0.1 to 3 μm and for UF from 0.01 to 0.1 μm. Otherwise, variations in membrane technologies are almost unlimited: membrane

Figure 7.5 Membrane filtration processes

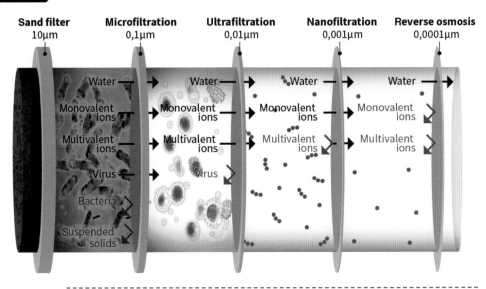

packaging, flow direction, material, pressure drop, etc. can vary greatly from one type to another.

Two main families of MF/UF membranes are generally used: ceramic and polymeric. Ceramic MF/UF membranes are made from oxides, carbides, or metals. They are mechanically strong, they are chemically and thermally stable, and they can achieve high flux rates. This type of membrane can remove particles, oil, oxides, and organic matter. They normally perform better than polymeric membranes at removing residual oil in produced water but do require frequent backwash.

Polymeric membranes are generally quite inexpensive but also more fragile than ceramic membranes. Integrity testing is required to ensure that the membrane is not damaged and operates properly. Typical materials used for their construction are polyvinylidene fluoride (PVDF) and polyacrylonitrile (PAN).

Published and unpublished tests have shown that it is possible to treat produced water containing polymers without significant impairment of the membrane efficiency. The choice of pore size is paramount to maintain deoiling efficiency while allowing the polymer to pass through and be disposed of or reused [12].

7.5. Polymer Degradation

Depending on the polymer concentration, the viscosity of the produced effluents, and the water treatment facilities in place, it might be necessary in some cases to degrade the polymer to maintain acceptable cleaning performance.

To begin with, a few questions should be addressed. The first is linked to the characteristics of the produced polymer. Determining average parameters may require running experiments in the laboratory with a model polymer. In practice, the produced polymer will have a lower molecular weight and a higher degree of hydrolysis. Therefore, the native polymer should not be used to quantify the effects on water treatment facilities.

Some published field cases in China have shown that, on average, molecular weight, concentration, and viscosity were 50–80% lower than injected, with an increase of hydrolysis degree from 5% to 25%. As discussed in the previous section, this aspect is linked to both the injection strategy (required pore volume for injection) and degradation occurring in the production facilities (pumps, air exposure, sampling strategy).

The second question is, can the polymer be reused? Even with a lower molecular weight, the produced polymer can be reused to dissolve new polymer powder, generating a fresh solution, with a possible dosage decrease in the long term and therefore significant savings. This aspect should be investigated when selecting a water treatment process.

The produced polymer can have several benefits [16]:

- It can be reused to possibly decrease the amount of fresh polymer needed to reach the target viscosity, thus reducing operating expenses (OPEX). This has no impact on standard dissolution facilities.

- Any polymer remaining in the water will act as a friction reducer water is pumped for injection or disposal, thereby reducing the pumping energy required and increasing injection rates for the same pressure drop.

- If water is injected into a virgin part of the reservoir, the polymer in the water phase will act as a sacrificial agent by adsorbing onto the rock. Therefore, if chemical injection is envisaged, fewer chemicals will be lost by retention.

- Anionic polymers have other benefits, such as acting as clay stabilizers, sand control, and even anti-corrosives.

In order to maintain good water treatment efficiency, it is possible to chemically or mechanically degrade the polymer or even remove it. Several methods can be considered [17-20]. A brief description will be given in the coming sections. However, not all are applicable in the field: it depends on local constraints, flow rates, and other characteristics of the effluents. A case-by-case approach must be used to find the most appropriate solution. Potential methods are as follows:

- Polymer precipitation with a coagulant
- Chemical degradation with an oxidizer
- Electro-oxidation
- Mechanical degradation
- Ultrasonic degradation
- Thermal degradation
- UV – advanced oxidation process (Fenton reaction)

7.5.1. Polymer Removal

Removing polymers from previous effluents is possible but very costly and not applicable everywhere. In addition to requiring significant quantities of chemicals to react with the polymer, the creation and handling of large sludge volumes may render the process uneconomical [8]. The main techniques are as follows:

- Polymer flocculation with organic or inorganic cationic coagulants. The polymer precipitates along with the O/W and the solids. The dosage required is on a 1 : 1 basis.

- Electro-coagulation. This process doesn't require chemicals but was not tested at large scales. Regular maintenance is needed to maintain acceptable removal efficiency.

- Precipitation with particles. A polymer is a flocculant. The addition of clays such as kaolinite will therefore remove the polymer from water. However, the dosage required to remove all polymers is prohibitive: 200 ppm of kaolinite are required to remove 1 ppm of polymer, and the sludge generated should be handled and treated.

7.5.2. Chemical Oxidation

The generation of free radicals can break the polymer backbone, as discussed in Chapter 4. Common oxidizers used in the industry are sodium hypochlorite (NaClO – bleach), hydrogen peroxide, sodium persulfate, and magnesium peroxide. Studies have shown that bleach is the most effective oxidizer [11, 21]. Laboratory studies have shown that 200 ppm of NaClO are enough to reach a viscosity of 1 cP in less than 10 minutes (Figure 7.6). There is the possibility of generating bleach on-site with electro-chlorination units. Residual chlorine should be removed before dissolving fresh polymer, to avoid degradation. Excessive chlorine consumption can occur if other reducers are present in the water, such as H_2S.

7.5.3. Electro-Oxidation

Electrodes can be used to generate a current that will help create hydroxy radicals to attack the polymer. Even though it has proven to be efficient at breaking viscosity, this approach also oxidizes the oil, creates foam, and leaves too many free radicals in the treated water.

Figure 7.6 ## Chemical degradation of a polymer solution with hypochlorite in the presence or absence of H₂S

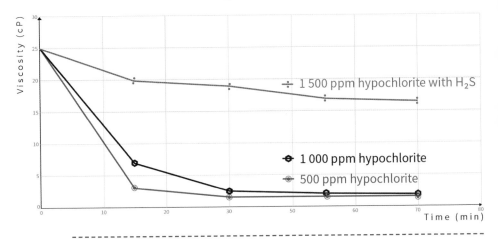

Figure 7.6

Chemical degradation of a polymer solution with hypochlorite in the presence or absence of H_2S

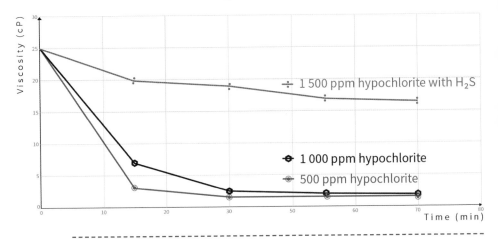

7.5.4. Mechanical Degradation

Breaking polymer chains will rapidly lead to decreased viscosity. This can be achieved using constrictions found in valves, pumps, chokes, or strong agitation. Applying a pressure drop above 40 bar will dramatically affect the polymer size and solution viscosity (Figure 7.7).

7.5.5. Ultrasonic Degradation

The use of ultrasound technology can degrade the polymer and allow a reduction in viscosity in a couple of minutes [12] with powers from 120 to 160 W. This method is not affected by the presence of contaminants. However, no large units capable of handling oilfield flow rates are available, and significant power consumption might be required to obtain a timely result.

Mechanical degradation of two polymer solutions through a valve

Figure 7.7

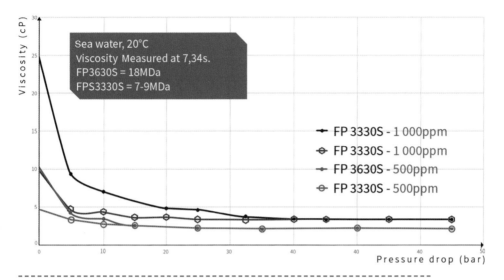

Sea water, 20°C
Viscosity Measured at 7,34s.
FP3630S = 18MDa
FPS3330S = 7-9MDa

- FP 3330S - 1 000ppm
- FP 3330S - 1 000ppm
- FP 3630S - 500ppm
- FP 3330S - 500ppm

Pressure drop (bar)

7.5.6. Thermal Degradation

It is possible to consider degrading polymers through heated tubes with a transit time of a couple of minutes (at temperatures above 180 °C). No chemicals are needed; but, on the other hand, scaling and corrosion issues should be expected for field brines with problematic compositions, in addition to very high energy consumption.

7.5.7. UV – Advanced Oxidation Processes

This technique uses UV (photolysis) or a combination of UV and radicals (photo-Fenton). It is very effective at reducing viscosity in less than five minutes but requires significant power and large facilities to treat the volumes of effluents produced in oil and gas fields. The presence of other reducing agents, such as H_2S, will also impact the efficiency of the process [22].

7.6. Conclusions and Discussion

Predicting the concentration and characteristics of the produced polymer is a difficult task. It will largely depend on the injection strategy and the equipment at the injection and production sides, which can change the molecular weight of the product injected. Moreover, transit through the reservoir can affect the hydrolysis and chemical size and composition of the product. Finally, dilution through commingled production can greatly impact the final viscosity arriving in the water treatment facilities. It is very difficult to reproduce in the laboratory typical effluents with oil, solids, and other contaminants. Care should therefore be taken when trying to model the possible impact of produced polymers on treating facilities.

In summary:

- Chemical and mechanical degradations are the most efficient techniques to reduce polymer viscosity. In particular, mechanical degradation can be performed inline with simple valves, at low cost; the main risk is the formation of more stable emulsions.

- Chemical degradation requires additional equipment and chemicals and is likely not available for offshore applications. A viscosity loss of 50% is enough to maintain good treatment efficiency in most cases; it can be achieved with a 50 bar pressure drop.

- Changing the demulsifiers/deoilers will likely be required if the polymer breaks through. New formulations exist to tackle this potential issue. A close collaboration with chemical suppliers is therefore mandatory.

- Gas flotation, hydrocyclones, and sand filters are affected the most by the presence of polymers, with deoiling efficiencies decreased by more than 50%.

- Walnut-shell filters, membranes, and other proprietary media filters are the most promising technologies, especially when combined with coalescers (Table 7.2).

Table 7.2

Summary of existing water treatment technologies and their relative efficiency in treating produced water containing polymers

Water treatment technology	Relative efficiency with polymer	Remarks
Nutshell filter		
Sand filter		
Siemens PerforMedia		Proprietary technology
Schlumberger MYCELX		Proprietary technology
Membranes		Choice of pore size is paramount
Centrifugation		
Hydrocyclones		
DGF		
IGF		
Flotation		Depends on the polymer viscosity
Gravity separation		Depends on the polymer viscosity
Coalescence		Depends on the polymer viscosity and concentration

(red = inappropriate/inefficient, green = appropriate/efficient, with relative efficiency from left [0] to right [maximum])

Combinations of technologies might be required to efficiently treat effluents containing polymers: a coalescer with nutshell filters or other media filters for example.

The use of membranes can help reuse the polymer, thus decreasing the need for fresh product and allowing significant savings in OPEX.

To precisely analyze the polymer produced, it is also necessary to implement specific protective sampling techniques to ensure that no further change in polymer characteristics occurs.

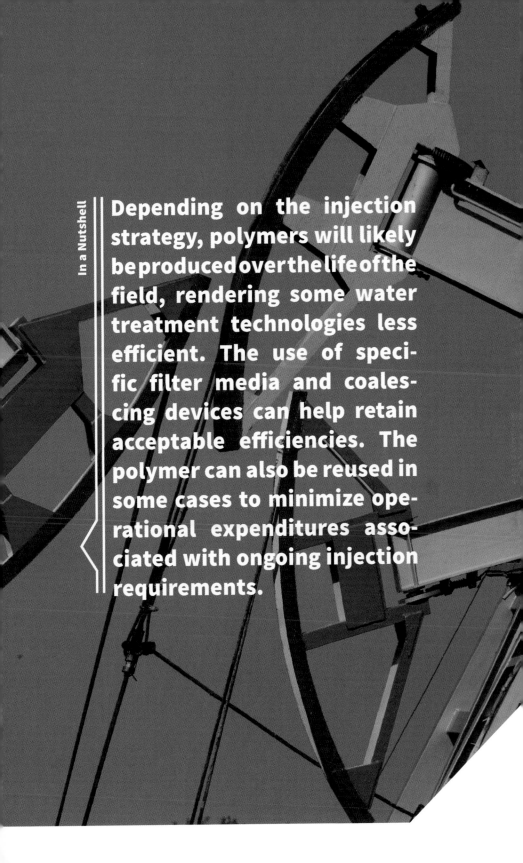

In a Nutshell

Depending on the injection strategy, polymers will likely be produced over the life of the field, rendering some water treatment technologies less efficient. The use of specific filter media and coalescing devices can help retain acceptable efficiencies. The polymer can also be reused in some cases to minimize operational expenditures associated with ongoing injection requirements.

References

[1] Lake, L.W. (2007). *Petroleum Engineering Handbook*, vol. III & IV. SPE. ISBN: 978-1-55563-126-0.

[2] Deng, S., Bai, R., Chen, J.P. et al. (2002). Produced water from polymer flooding process in crude oil extraction: characterization and treatment by a novel crossflow oil-water separator. *Separation and Purification Technology* 29: 207–216. Elsevier.

[3] Di, W., Guangzhi, A., Yan, C. et al. 1999. Rheology and stability of ior produced liquid in daqing oilfield. Paper SPE57318 presented at the SPE Asia Pacific Improved Oil Recovery Conference, Kuala Lumpur, Malaysia, 25–26 October.

[4] Bartz, D. and Gotterba, J. 2014. Results of field operation of a distributed flux burner in a heater treater in a northern Canada heavy oil field; thermal performance and firetube life. Paper SPE170172 presented at the SPE Heavy Oil Conference, Alberta, Canada, 10–12 June. https://doi.org/10.2118/170172-MS.

[5] Wylde, J.J., Allan, K., McMahon, J. et al. 2011. Scale inhibitor application in northern Alberta – a case history of an ultra high temperature scale inhibition solution in fire tube heater treaters. Paper SPE141100 presented at the SPE International Symposium in Oilfield Chemistry, The Woodlands, Texas, USA, 11–13 April. https://doi.org/10.2118/141100-MS.

[6] Al-Kalbani, H., Mandhari, M.S., Al-Hadhrami, H. et al. 2014. Impact on crude dehydration due to back production of polymer. Paper SPE169718 presented at the SPE EOR Conference at Oil & Gas West Asia, Muscat, Oman, 31 March – 2 April. https://doi.org/10.2118/169718-MS.

[7] Zheng, F., Quiroga, P., Zaouk, M. et al. 2013. Electrostatic dehydration of heavy oil from polymer flood with partially hydrolyzed polyacrylamide. Paper SPE 165269 presented at the SPE Enhanced Oil Recovery Conference, Kuala Lumpur, Malaysia, 2–4 July. https://doi.org/10.2118/165269-MS.

[8] Fei, Z.X., Xin, L.L., Chan, W.Y. et al. (2007). Influence of residual polyacrylamide mass concentration upon waste water treatment with polymer flooding flocculation. *Journal of Daqing Petroleum Institute* 31.

[9] Al-Maamari, R.S., Sueyoshi, M., Tasaki, M. et al. 2014. Polymer-flood produced-water-treatment trials. Paper SPE172024 presented at the Abu Dhabi International Petroleum Exhibition and Conference, Abu Dhabi, UAE, 10–13 November. https://doi.org/10.2118/172024-MS.

[10] Kaiser, A., White, A., Lukman, A. et al. 2015. The influence of chemical EOR on produced water separation and quality. Paper SPE 174659 presented at the SOE Enhanced Oil Recovery Conference, Kuala Lumpur, Malaysia, 11–13 August. https://doi.org/10.2118/174659-MS.

[11] Rambeau, O., Alves, M-H., Loriau, M. et al. 2015. Chemical solutions to handle viscosified back produced water in case of polymer flooding. Paper SPE177501

presented at the Abu Dhabi International Petroleum Exhibition and Conference, Abu Dhabi, UAE, 9–12 November. https://doi.org/10.2118/177501-MS.

[12] Rambeau, O., Alves, M-H., Andreu, N. et al. 2016. Management of viscosity of the back produced viscosified water. Paper SPE 179776 presented at the SPE EOR Conference at Oil & Gas West Asia, Muscat, Oman, 21–23 March. https://doi.org/10.2118/179776-MS.

[13] Ma, H., Abdullah, S., Shawabkeh, R. et al. 2017. Destabilization and treatment of produced water-oil emulsions using anionic polyacrylamide. Paper SPE183665 presented at the SPE Middle East Oil & Gas Show and Conference, Manama, Kingdom of Bahrain, 6–9 March. https://doi.org/10.2118/183665-MS.

[14] Richerand, F. and Peymani, Y. 2015. Improving flotation methods to treat eor polymer rich produced water. Paper SPE174535 presented at the SPE Produced Water Handling and Management Symposium, Galveston, Texas, USA, 20–21 May. https://doi.org/10.2118/174535-MS.

[15] Hendricks, D. (2006). *Water Treatment Unit Processes: Physical and Chemical.* CRC Press – Taylor & Francis Group.

[16] Thomas, A. (2016). Polymer flooding. In: *Chemical Enhanced Oil Recovery (cEOR) – a Practical Overview* (ed. L. Romero-Zerón). InTech https://doi.org/10.5772/64623.

[17] Guangmeng, R., Dezhi, S., and Meiling, W. (2006). Progress in the treatment technologies for wastewater from tertiary oil recovery in China. *Industrial Water Treatment* 26.

[18] Go-Yin, Y., Shan, G.L., Zhou, T.S. et al. (2006). Treating recycled oilfield produced water treatment technology by electrochemical oxidation/coagulation for corrosiveness reduction. *Oilfield Chemistry* 23.

[19] Xueguang, Z., Wu, C., Ping, M., and Yangqiu, P. (2009). Advance on treatment technology for oil-gas field operation flowback fluids in China. *Advances in Fine Petrochemicals* 10.

[20] Wang, Z., Lin, B., Sha, G. et al. 2011. A combination of biodegradation and microfiltration for removal of oil and suspended solids from polymer-containing produced water. Paper SPE140916 presented at the SPE America E&P Health, Safety, Security and Environmental Conference, Houston, Texas, USA, 21–23 April. https://doi.org/10.2118/140916-MS.

[21] Thomas, A., Gaillard, N., and Favéro, C. (2013). Some key features to consider when studying acrylamide-based polymers for chemical enhanced oil recovery. *Oil & Gas Science and Technology – Rev. IFP Energies Nouvelles* https://doi.org/10.2516/ogst2012065.

[22] Qiang, S., Guangxu, Y., and Shaohui, G. (2007). Experimental study on removal of polyacrylamide in the oilfield polymer-bearing wastewater. *Energy Environmental Pollution* 21.

Chapter eight

Economics

Economics of chemical enhanced oil recovery techniques are critical to the implementation and the success of any project. Some basic calculations and considerations will be addressed in this chapter.

Essentials of Polymer Flooding Technique, First Edition. Antoine Thomas.

8.1. Introduction

In chemical enhanced oil recovery (EOR) processes, the main expenses are always associated with the chemicals injected (operational expenditures [OPEX]). Capital expenditures [CAPEX] linked to surface facilities usually account for 10–20% of the budget, while costs linked to laboratory studies and the design phase represent a maximum of 1% over the life of a full commercial project. Obviously, the more components included in the chemical slug (from P, to surfactant-polymer [SP] and alkali-surfactant-polymer [ASP]), the higher the costs; but, in theory, additional chemicals are included to ensure greater ultimate oil recovery.

An important factor dictating the choice of one technique over another is the oil field location. Given the requirements for ASP, for instance (water softening, logistics, dissolution facilities), there is a high probability that such processes will never be implemented at large scales offshore – not to mention the need for disposing or reusing the produced water containing the various components.

Every oil field has, by definition, its own characteristics. Every country has a specific tax regime. It is almost impossible to provide a precise economic evaluation without knowing the primary expenses specific to each jurisdiction. Therefore, in this chapter, we will examine simplified examples of the OPEX and CAPEX linked to the EOR process. The idea is to outline the main contours of a typical chemical EOR approach and give an order of magnitude for the associated costs.

8.2. Cost Overview

The following list summarizes the primary expenses incurred when deploying a chemical EOR technology [1–6]:

- Well workovers or infill drilling required to optimize the spacing and completion of a pilot test.

- Water injection and handling.

- *Water treatment facilities* for separation and water treatment. Some changes might be required, depending on the formulation injected and the existing facilities.

- *Water softening for systems that use alkali.* Required for ASP processes to remove divalent cations.

- *EOR chemicals.* Polymers with or without surfactants and alkali.

- *Other chemicals.* Scale inhibitors, deoilers, emulsion breakers, corrosion inhibitors, biocide, etc.

- Chemical storage facilities for chemicals.

- Storage tanks for softened water.

- Tanks for waste storage and disposal.

- Dissolution, mixing, and injection facilities.

- Onsite laboratory.

- Logistics.

- Access roads.

- Manpower and OPEX.

Some examples of cost splits will be detailed in the following sections.

8.2.1. Water Handling and Injection

Water-handling costs are very important to consider when injecting polymers in a brownfield where the water-cut has reached its economic limit. Polymer injection will indeed, in the vast majority of cases, induce a water-cut decrease, which can be significant depending on the reservoir history and presence of an aquifer. For instance, in some patterns of the Daqing oil field, a 10% decrease in water-cut was observed, translating into major savings [7]. In that particular case, the analyses concluded that the cost of polymer flooding was $0.65/barrel lower than for waterflooding.

The typical divisions for water-handling stages are lifting, separating, deoiling, filtering, pumping, and injecting. In an oil field review paper [8], typical costs for waterflooding were documented and have been summarized in Table 8.1; the split by cost category is indicated in Table 8.2.

In these categories, the addition of chemicals represents approximately 15%, while the surface facilities represent 55% of

Table 8.1

Example of water injection costs

Item		Cost ($/bbl)
Water injection		0.03
Water lift		0.04
Total cost (including lifting, separation, deoiling, filtering, pumping and injection, chemicals, and utilities)	$20\,000\,\mathrm{bbl\,d^{-1}}$	0.842
	$50\,000\,\mathrm{bbl\,d^{-1}}$	0.559
	$100\,000\,\mathrm{bbl\,d^{-1}}$	0.478
	$200\,000\,\mathrm{bbl\,d^{-1}}$	0.434

Table 8.2

Cost split by category for waterflooding

Item	Percentage of expenses (average, %)
Lifting	18
Separation	15
Deoiling	20
Filtering	15
Pumping	27
Injecting	5

overall CAPEX. It is important to review these material and infrastructure costs for each project when switching from water-flood to polymer flood; however, it is also important to include the potential savings, especially for lifting, pumping, and injection. A polymer can act as a friction reducer, reducing the pressure drop by up to 70% in pipes where the flow is turbulent. Given electrical requirements ranging from 0.03 to 2 kw/bbl, this could result in significant savings for full-field applications.

8.2.2. Cost of EOR Chemicals

Over the past 30 years, many improvements in manufacturing processes and chemistry have led to a decrease in the polymer cost per concentration (ppm) injected. The following are some of these improvements:

- Industrial manufacturing plants have larger reactor capacities.
- More stable chemistries have been developed.
- Higher molecular weights mean less polymer is required to reach a target viscosity.
- Optimized surface facilities and monitoring strategies help minimize polymer degradation.

For surfactants, the equation has more unknowns. A formulation – and, sometimes, brand-new molecules – must be developed for each unique reservoir and crude oil. This implies the use of specific raw materials and the construction of suitable manufacturing capabilities to supply products in a timely fashion and at a reasonable cost.

The typical range of costs, not including logistics (ex works), are shown in Table 8.3:

In addition, cosolvents (alcohol) are sometimes included in SP and ASP formulations to enhance the performance of surfactants.

Table 8.3

Typical range of costs for the chemicals used in enhanced oil recovery processes

Chemical	Cost ($ kg⁻¹)	Typical dosages (% active)
Polymer	2.5–5.5	0.1–0.3
Surfactant	5–15	0.5–1
Alkali	0.25–0.45 for Na_2CO_3 0.55–1.10 for NaOH (100% active)	0.5–1.5

Calculating the cost per barrel injected is simple. Given a polymer flood and assuming 1000-ppm active polymer injected at $3/kg, we obtain:

$$\textit{Polymer cost} = 1000\,(\textit{ppm}) * 10^{-6} * 3\,(\$/kg) * 158.75\,(kg/bbl)$$
$$= \$0.47/bbl\ injected$$

If the reservoir pore volume and the concentration of polymer to be injected are known, it is possible to calculate the total polymer expenses for the particular field. Examples will be given in upcoming sections.

8.2.3. Additional Costs for ASP Flooding

In ASP processes, it is necessary to soften the injection water to remove divalent cations that would cause precipitation of alkali components. Depending on the salinity and the volume of water to treat, the costs can be quite high. For instance, the cost per barrel to soften water usually ranges from $0.015/bbl for strong acid cation (SAC) resins to $0.35/bbl for weak acid cation (WAC) resins. Given the need for water-softening units, the plant costs for an ASP facility is usually three times that for SP processes and four times the facility requirement for polymer flooding. This is especially true when only saline water is available in the field to prepare the chemical blend. Also, an important CAPEX in ASP plants is linked to the oil/water treatment facilities required to break the emulsion and separate the oil, while treating the produced water.

8.3. Example – Polymer Flooding

In this example, we consider a sandstone reservoir and, more precisely, a five-spot pattern in a confined area of the field with the characteristics outlined in Table 8.4.

Characteristics of the five-spot pattern used in the calculations

Table 8.4

Parameter	Definition/Value
Lithology	Sandstone
Reservoir thickness	50/15.2 ft m^{-1}
Area (acre ha^{-1})	50/20.2 acre ha^{-1}
Porosity (%)	~25.7%
Pattern gross pore volume (bbl)	5 000 000 bbl
Initial oil saturation, fraction of PV	0.75
Original-oil-in-place (bbl)	3 750 000 bbl
Recovery factor after waterflood (%OOIP)	35%
Oil volume remaining after waterflood (bbl)	3 437 500 bbl
Oil saturation after waterflood (fraction of PV)	0.49

We assume that, after 20 years of production and water injection, only 35% of oil originally in place (OOIP) has been recovered, leaving 1 314 563 bbl of oil in the reservoir. The water-cut in the producer has reached 98% and is likely very close to the economic limit. In this case, chemical EOR methods must be considered to further increase the final recovery factor.

After careful reservoir and fluid analysis, including consideration of both oil viscosity and reservoir heterogeneity, a viscosity value was chosen to obtain a mobility ratio value below 1. Laboratory tests were performed to select the best polymer candidate in representative injection brine to reach the target viscosity. In this case, water salinity and reservoir temperature were on the low side of the range, and the target viscosity was reached with 1500-ppm active product, utilizing a conventional copolymer of acrylamide and acrylic acid with high molecular weight. Table 8.5 summarizes the design for the pilot:

Table 8.5 **Summary of injection parameters for the five-spot pattern – polymer injection**

Parameter	Value
Total injection rate	5000 bpd
Polymer active concentration	1500 ppm
Polymer cost	$3 kg^{-1}
Reservoir pore volume injected	50%
Void replacement ratio	1
Volumetric sweep efficiency (for EOR)	0.8
Time to inject 50% PV	16.4 mo
Total fluid rate (production)	5000 bpd
Water-cut (before EOR)	98%
Total water rate	4900 bpd
Total oil rate	100 bpd
EOR	
Polymer cost per bbl	$0.71/bbl injected
Polymer cost per day	$3550
Polymer required (50% PV)	382 tons
Total polymer cost (50% PV)	$1144059
Total chemical costs	$1144059

In this case, we assume no injectivity change between water and polymer. However, in real life, it is common for injectivity to decline with continued polymer injection; therefore, the time to fill 50% of the reservoir pore volume would obviously change, as would the net present value (NPV). Also, no change in void-age replacement ratio (VRR) would be expected in that case.

Given an oil price of $45/bbl and a working interest of 100%, we can calculate several parameters, as illustrated in Table 8.6.

Table 8.6

Cost of recovery as a function of the percentage of extra oil recovered

Extra OOIP recovered (%)	1	3	5	7	10	15	20
Volume of oil (1000s of bbl)	39	113.1	189.2	263.2	375	563.5	75
Oil revenue (million $)	1.8	5.1	8.5	11.9	16.9	25.4	35.8
Chemical cost/bbl recovered ($/bbl)	29.3	10.1	6.1	4.4	3	2	1.5
Ratio of profit/ chemical cost	0.53	3.45	6.44	9.35	13.77	21.17	28.5

On average, polymer flooding helps recover an extra 10% OOIP, based on existing field cases and corresponding publications. In this particular case, looking at the 10% extra OOIP, the cost per barrel recovered is approximately $3/bbl, which is within the typical OPEX numbers obtained for such projects.

It is possible to add the cost of equipment over the injection period of 16.4 months, assuming a lease option (Table 8.7).

Table 8.7

Typical equipment-related costs

Parameter	Value
Injection period	16.4 months
Monthly leasing cost for equipment (all-inclusive)	$40 000
Total expenditures over the injection period	$656 000
Monthly injection rate	152 500 bbl
Cost per bbl injected	$0.26

The total cost per barrel injected (polymer + equipment) for 16.4 months of injection is equal to $0.97/bbl injected. The total expenditures over that same period are equal to approximately $1.8 million. Breakeven point is reached with extra oil recovery of 1.1% OOIP (40 000 barrels of oil). Obviously, this does not include expenditures related to other infrastructures, manpower, etc. But it gives an order of magnitude of what can be expected when starting a polymer injection pilot. Several remarks can be made:

- Injection equipment usually has the capacity to inject into several wells (more than 10, depending on rates, concentration, and constraints). Therefore, only the addition of modules may be required for extension, limiting the extra CAPEX.

- The initial cost of skid-mounted equipment is generally amortized over several years. After this period, the $40 000/month will no longer be a factor. The remaining costs will be linked to maintenance.

It is interesting to compare the cost of a pilot ($1.8 million total, in this case) with the cost of drilling a well onshore. Given drilling costs from $0.5 to $4 million and the risk associated with the absence of discovery or a dry well, a polymer flood is a good alternative to produce more oil, at lower risk and lower cost.

The idea after such a pilot, if successful, is to continue with the same piece of equipment and add modules with pumps and, possibly, other dissolution equipment to handle more wells. This strategy is very complementary to phased development and implementation in larger fields.

The main conclusion that can be drawn from this rough economic calculation is that, even at the pilot stage, it is possible to obtain positive results both technically and economically if the approach to implementation is appropriately designed.

8.4. Examples – SP and ASP

8.4.1. SP

For this example, we will consider the same field as in Section 8.3 and a pattern with the exact same characteristics (injection over 16.4 months). In addition to injecting a polymer at 1500-ppm active concentration, a surfactant is added with an active concentration of 1% by weight (including all components). A summary of the injection parameters is given in Table 8.8.

Summary of injection parameters – surfactant-polymer injection

Table 8.8

Parameter	Value
Total injection rate	5 000 bpd
Polymer active concentration	1 500 ppm (0.15 wt%)
Polymer cost	$3 kg^{-1}
Surfactant active concentration	1 wt%
Surfactant cost	$7.5 kg^{-1}
Reservoir pore volume injected	50%
Volumetric sweep efficiency (for EOR)	0.8
EOR	
Polymer cost per bbl	$0.71/bbl injected
Polymer cost per day	$3 550
Polymer required (50% PV)	382 tons
Total polymer cost (50% PV)	$1 144 059
Surfactant cost per bbl	$11.92/bbl injected
Surfactant cost per day	$59 600
Surfactant required (50% PV)	3 180 tons
Total surfactant cost (50% PV)	$23 869 619
Total chemical costs	$25 013 678

The cost of equipment is not very different between a P and a SP: storage is required for the surfactant along with a metering pump to dose the product in the flow line and a mixer to homogenize the solution.

In this case, just considering the OPEX, breakeven is reached with extra oil production of approximately 556 000 barrels or ~15% OOIP.

Obviously, the main expense is the surfactant, with the cost/ bbl injected increasing by a factor of nearly 17 between P and SP, given the chemical costs and concentrations injected.

8.4.2. ASP

Let's again consider the same reservoir, but this time for an ASP flood. The main objective of including alkali in the formulation is to decrease the surfactant concentration necessary to reach the optimum interfacial tension and therefore decrease surfactant chemical consumption. For the sake of simplicity, we'll consider the injection over the same pore volume (50%, taking 16.4 months) with a 30% ASP slug pushed by a 20% polymer slug. The results are summarized in Table 8.9.

Table 8.9

Summary of injection parameters – alkali-surfactant-polymer injection

Parameter	Value
Total injection rate	5 000 bpd
Polymer active concentration	1 500 ppm (0.15 wt%)
Polymer cost	$3 kg^{-1}
Surfactant active concentration	0.5 wt%
Surfactant cost	$7.5 kg^{-1}
Alkali active concentration	1 wt%
Alkali cost	$0.5 kg^{-1}
Reservoir pore volume injected	30% ASP + 20% P
Volumetric sweep efficiency (for EOR)	0.8
EOR	
Polymer cost per bbl	$0.71/bbl injected
Polymer cost per day	$3 550
Polymer required (50% PV)	382 tons
Total polymer cost (50% PV)	$1 144 059

...

Parameter	Value
Surfactant cost per bbl	$5.96/bbl injected
Surfactant cost per day	$29 800
Surfactant required (30% PV)	954 tons
Total surfactant cost (30% PV)	$7 160 886
Alkali cost per bbl	$0.81/bbl
Alkali cost per day	$4 050
Alkali required (30% PV)	4 770 tons
Total alkali cost	$2 418 508
Total chemical costs	$10 723 452

For the case of the ASP flood, the breakeven point is reached with extra oil recovery of approximately 6.5% OOIP or 240 000 barrels.

The cost of equipment is higher for ASP compared to SP, since water-softening units are required. Assuming, for example, $0.10/bbl of water softened using a resin, this would add $250 000 over the lifetime of the project (16.4 months), not including the CAPEX for the water treatment facilities. Given that the bulk of the expenses are related to the surfactant, it is logical that, from an OPEX standpoint, ASP appears cheaper than SP. However, in existing field cases, designs often vary between P, SP, and ASP. For instance, it is not unusual to see an ASP with the following characteristics:

- *Injection of an ASP slug*: 40% of reservoir pore volume.

- *Injection of a polymer slug following the ASP slug*: 30% of reservoir pore volume.

- *Polymer dosage*: 2000 ppm (0.2 wt%) active.

- *Surfactant + co-solvent dosages*: 5000 ppm (0.5 wt%) active.

- *Alkali dosage*: 10000 ppm (1 wt%) active.

In addition, it should be remembered that the surface facilities for ASP are up to three times more expensive than for SP.

In some cases, the use of sodium hydroxide as alkali has led to major scaling issues, with numerous pump failures at the production side. Recently, ammonia has been reconsidered as a potential alkali for ASP.

8.4.3. Comparison P – SP – ASP

From the literature, it appears that average chemical concentrations for P, SP, and ASP are as follows:

- *Active polymer concentrations*: from 1000–3000 ppm (0.1–0.3 wt%).

- *Active surfactant concentrations*: from 1500 to 7000 ppm (0.15–0.7 wt%) for ASP, and from 7500 to 15000 ppm (0.75–1.5 wt%) for SP.

- *Active alkali concentrations*: from 5000 to 15000 ppm (0.5–1.5 wt%).

Table 8.10 summarizes typical OPEX and other figures related to the implementation of chemical EOR in a shallow sandstone reservoir with low salinity and temperature:

When all three methods are compared, from an OPEX standpoint, ASP and SP are 10.5 and 17.8 times more expensive than polymer flooding, respectively. Theoretically, ASP is supposed to increase the recovery factor by 20% OOIP or more; this supposes careful laboratory and field designs to ensure that the formulation can propagate without too much retention, while decreasing the interfacial tension to sufficiently low values to allow residual oil to be mobilized and recovered.

Among the other critical aspects are the following:

- *Water treatment facilities.* For ASP, softening units are required, adding substantial costs to the project. Also, in SP and ASP, tight emulsions can be formed, requiring specific chemicals and treatment strategies to separate oil from water and treat the produced water.

- *Logistics.* The more chemicals, the higher the transportation costs and storage required. For an ASP pilot with

Comparison of relative costs for P, SP, and ASP injections

Table 8.10

	P	SP	ASP
Polymer cost		$3 kg^{-1}	
Polymer dosage		0.15 wt% active	
Polymer cost per bbl injected		$0.71/bbl	
Surfactant cost		$7.5 kg^{-1}	
Surfactant dosage	—	1 wt% active	0.5 wt% active
Surfactant cost per bbl injected	—	$11.92/bbl	$5.96/bbl
Alkali cost	$0.5 kg^{-1}		
Alkali dosage	—	—	1 wt% active
Alkali cost per bbl injected	—	—	$0.81/bbl
Total cost per bbl injected	$0.71/bbl	$12.63/bbl	$7.48/bbl
Average oil recovery	10% OOIP	15–20% OOIP	20–25% OOIP
Cost ratio to P	—	17.8	10.5

0.15 wt% active polymer, 0.5 wt% active surfactant, and 1 wt% active alkali with an injection rate of 5000 bpd, 13 tons of chemicals will be consumed every day – this increases to 52 tons for 20 000 bpd, 131 tons for 50 000 bpd, etc. However, the surfactants and alkali are not usually delivered under 100% active form, substantially increasing the need for efficient transportation. Logistics can be a real showstopper for full-field ASP developments. Progressive deployment is probably the most efficient way to minimize risks and issues.

8.5. Conclusions

In chemical EOR processes, OPEX are the largest components of the overall cost of implementation. Given the current costs of polymers, surfactants, and alkali with the typical concentrations used in these processes, we can state that:

- ASP and SP are, on average, from 10 to 15 times more expensive than P ($ per barrel injected), respectively. The final chemical cost will depend on the pore volume injected for the slugs and to what end retention may or may not impact this. In an ASP process with a polymer chase, the polymer can represent a significant expense.

- The main expense in SP and ASP is the surfactant.

- The surface facilities for ASP are usually three times more expensive than for SP. There is a little difference between P and SP from a CAPEX standpoint.

- Specific water-softening and treatment units are required for ASP to remove divalent cations and separate oil from water. This adds substantial costs to the whole project.

- Other aspects should be considered in ASP, such as scaling issues, logistics, selection of new deoilers and emulsion breakers, etc.

- The final cost per barrel recovered usually ranges from $7 to $10/bbl for P, $15 to $25/bbl for SP, and $20 to $35/bbl for ASP, including OPEX and CAPEX for all three cases.

- Typical average recovery factors are 10% OOIP for P, 15–20% OOIP for SP, and 20–30% OOIP for ASP.

The degree of difficulty obviously increases from P to SP and ASP, from both laboratory design and facility standpoints.

In a Nutshell

The chemical EOR methods of ASP and SP are, on average, from 10 to 15 times more expensive than P ($ per barrel injected), respectively. The primary expense in SP and ASP is the surfactant. The final chemical cost will depend on the pore volume injected for the slugs. In the ASP process with a polymer chase, the polymer can represent a significant expense.

References

[1] Sheng, J. (2013). A comprehensive review of ASP flooding. Paper SPE165358 presented at the SPE Western Regional & AAPG Pacific Section Meeting, 2013 Joint Technical Conference, Monterey, California, USA, 19–25 April. https://doi.org/10.2118/165358-MS.

[2] Manrique, E., De Carvajal, G., Anselmi, L. et al. (2000). Alkali/surfactant/polymer at VLA 6/9/21 field in Maracaibo Lake: experimental results and pilot project design Paper SPE 59363, presented at the SPE/DOE Improved Oil Recovery Symposium, Tulsa, Oklahoma, USA, 3–5 April. https://doi.org/10.2118/59363-MS.

[3] McInnis, L., Hunter, K., Ellis-Toddington, T. et al. (2013). Case study of the Mannville B ASP flood. Paper SPE 165264 presented at the SPE Enhanced Oil Recovery Conference, Kuala Lumpur, Malaysia, 2–4 July. https://doi.org/10.2118/165264-MS.

[4] Pandey, A., Koduru, N., Stanley, M. et al. (2016). Results of ASP pilot in Mangala field: a success story. Paper SPE179700 presented at the SPE Improved Oil Recovery Conference, Tulsa, Oklahoma, USA, 11–13 April. https://doi.org/10.2118/179700-MS.

[5] Sheng, J.J. (2011). *Modern Chemical Enhanced Oil Recovery: Theory and Practice*. Elsevier.

[6] Sheng, J.J. (2013). ASP fundamentals and field cases outside China, Chapter 9. In: *EOR Field Case Studies* (ed. J.J. Sheng), 189–201. Elsevier.

[7] Demin, W., Jiecheng, C., Junzheng, W. et al. (2002). Experiences learned after production of more than 300 million barrels of oil by polymer flooding in Daqing oil field. Paper SPE77693 presented at the SPE Annual Technical Conference and Exhibition, San Antonio, Texas, USA, 29 September – 2 October. https://doi.org/10.2118/77693-MS.

[8] Bailey, B., Crabtree, M., Tyrie, J. et al. (2000). *Water Control*. Spring: Schlumberger Oilfield Review.

Chapter nine

Field Cases

In this chapter, we will discuss some interesting aspects of a few key field cases using polymer, surfactant–polymer, and alkali-surfactant-polymer flooding (P, SP, ASP). There have been – and still are – many ongoing polymer floods around the world. Reviewing them all would be an arduous and time-consuming exercise and would not provide much additional knowledge. Every case is different, especially since limited information is made public; even though some comparisons can be drawn, it is always better to build a business case with ground science and common sense, rather than just copying what has been done in the past. It is important to build on experience and, more importantly, to keep an open mind and a fresh eye on all technological developments available for implementation. An old precept says that if you give a man a fish, you feed him for a day, but if you teach a man to fish, you feed him for life. So, let's build (on) knowledge.

Essentials of Polymer Flooding Technique, First Edition. Antoine Thomas.
© 2019 John Wiley & Sons Ltd. Published 2019 by John Wiley & Sons Ltd.

9.1. Introduction

In 2016, more than 50 polymer floods were operating world-wide, not including projects still in the design phase or about ready for deployment in the field. Every reservoir has its own constraints, and the injection strategy should be adapted accordingly. Several papers summarize chemical enhanced oil recovery (EOR) projects throughout the world, with varying degree of detail [1-6]. A list of ASP field projects was given in Chapter 2.

As for polymer injection, Standnes and Skjevrak [6] have listed several causes of failures in the field, which are discussed next. Additional references can be found in their paper.

Average permeability in the field too low (7 mD). Recent laboratory studies have shown that it was possible to inject in very-low-permeability cores (below 3 mD). This was confirmed by some injectivity tests (unpublished) in a carbonate reservoir in the Middle East. Very-low-permeability reservoirs are not the best candidates for polymer injection. However, there are several ways to improve the propagation in such reservoirs:

- Purposely inject a polymer with a very low molecular weight (less than 3 million Da)

- Preshear the polymer to remove the high-molecular-weight portion and enhance propagation.

In both cases, it is important to maintain a certain viscosity and find a compromise to safeguard injectivity while still sufficiently mobilizing oil and conforming reservoir sweep.

Injected polymer concentration too low (213 ppm). When the salinity of the injection water is low, it is possible to obtain a very high viscosity with a low polymer concentration. In this case, it is therefore possible to more easily obtain a mobility ratio of 1 while minimizing operating expenses (OPEX). However, this strategy neglects a very important point, which is polymer retention.

Given a sandstone with a density of $2.65\,\mathrm{g\,cm^{-3}}$, 25% porosity, and a polymer injected at an active concentration of 213 ppm exhibiting a retention value of $35\,\mathrm{\mu g\,g^{-1}}$, we can use the following formula to determine the pore volume delay factor:

$$PV_{ret} = \left[\rho_{rock}\left(1-\varphi\right)/\varphi\right] * \left[R_{poly}/C_{poly}\right]$$

where ρ_{rock} = rock density ($\mathrm{g/cm^3}$), φ = porosity (%), R_{poly} = polymer retention value ($\mathrm{\mu g/g}$), and C_{poly} = injected polymer concentration (ppm)

The pore volume delay factor PV_{ret} represents the additional pore volume (PV) required above and beyond the target PV determined, in order to offset the denuded polymer bank resulting from the level of retention indicated in the formula. A 100% PV_{ret} suggests that double the injected PV would be required to account for retention. Intuitively, then, if the effective pore volume injected is too low and insufficiently offsets the retention, there is a high probability that no polymer (and therefore no viscosity) will propagate in the reservoir to affect reservoir sweep and displacement as designed. Besides, viscosity does not increase linearly with concentration, but rather exponentially. This means that doubling the concentration will more than double the viscosity after the *critical concentration*, C*. This also means that losing some polymer because of retention will have an important impact on viscosity. Therefore, even if a small concentration is required to reach a target viscosity, it is necessary to compensate for possible retention in the reservoir; and it is therefore recommended to inject a higher polymer concentration than the base case design.

Viscosity reduction due to mixing between injection and formation waters. Detecting mixing with formation water is quite complex. Many factors can explain the fact that a low polymer concentration is measured in the producing wells or after back production in an injection well:

- Polymer retention
- Dilution and mixing
- Insufficient pore volume injected
- Insufficient viscosity injected relative to the heterogeneity or oil viscosity

In all cases, a simple way to minimize dilution by mixing is to increase the injected viscosity. Also, the lower the water saturation initially present in the reservoir, the less mixing and less propensity for viscous fingering to occur to connected or established water saturation pathways; this is why it is important to start polymer injection early in the life of the field.

Too high a resistance factor causing an unacceptable injectivity decrease. Generally speaking, several reasons can explain this observation:

- Bad water quality.
- Injection of polymers with very high retention.
- Injection of self-associating polymers such as hydrophobic associative polymers.
- Injecting polymers of unsuitable size relative to the permeability distribution.

This aspect can normally be de-risked with core flooding.

Very-high-permeability contrast (>100). When the Dykstra-Parson coefficient is too high, there is a risk that the polymer enters only the very-high-permeability zones. A possible way to mitigate this issue is to start by injecting very-high-viscosity slugs. Another strategy consists of doing a conformance treatment with an appropriate technology to fill the offending thief zone.

Much higher polymer retention than expected. This risk is often associated with bad reservoir understanding or characterization.

Geological facies variations, zones with a higher percentage of clays, and large permeability changes can be the reasons for such observations. In the field, it is likely that polymer retention will be higher than in the laboratory (even with reservoir cores) because of the aforementioned uncertainties. It is possible to inject sacrificial agents or slightly increase the concentration of the chemicals injected to compensate for retention. With regard to sacrificial agents, it is necessary to calculate the quantity required to saturate the part of the reservoir that will be contacted later by the chemical enhanced oil recovery (CEOR) fluids and ensure that these same reservoir volumes will be contacted by both the sacrificial chemical and the CEOR slug.

9.2. Envelope of Application

In order to better define an envelope of application, instead of analyzing all field cases, it is possible to look at the extreme cases where a polymer has been injected.

Example 9.1 (Unpublished).
The field is located in Eastern Europe, with the primary field characteristics summarized in Table 9.1. Given the usual parameters used to screen candidates for polymer flooding, this field would not rank at the top. With a low oil viscosity, this clearly is not a good candidate from a mobility ratio perspective, not to mention the high-temperature salinity. However, only 29% of the oil originally in place (OOIP) was recovered over the years of exploitation with water-cut over 95% in all the producing wells.

The field is an anticline with peripheral injection and production at the top of the structure. A strong edge aquifer is present in the west part of the field, rendering the exploitation of nearby wells uneconomical. Well spacing is important, with average distances of 600 m between the injection and production wells.

Table 9.1 **Field characteristics of Eastern European field 1**

Property	Value and units
Lithology	Sandstone
Average permeability	400 mD
Injection water salinity	$80\,g\,l^{-1}$
Divalent cations	$6\,g\,l^{-1}$
Reservoir temperature	$80\,°C$
Oil viscosity	$0.4–0.9\,cP$

A tracer test was performed and showed very rapid break-through in the nearby wells (a couple of hours), confirming the existence of very-high-permeability streaks. Conformance treatments were considered at first; but, for a number of reasons that will not be detailed here, no well workovers or zone isolation were possible. It was deemed too risky to bullhead any treatment without zone isolation, since any fluid entering the low-permeability zones (based on Darcy's law) would greatly impair overall injectivity. Moreover, plugging such a thief zone would require a large volume of chemical to make sure water or polymer did not break through or quickly circumvent it.

It was decided to inject very-high-viscosity polymer slugs (150 cP) to displace the 0.5-cP oil and greatly decrease the intake of injectant by the high-permeability zones. Injection was performed in two wells, one to the west and one to the east of the structure, monitoring the central row of producers (six wells). Starting with such a high-viscosity slug prevented the aquifer from moving upward, also greatly minimizing dilution. We can picture a very stable displacement from the bottom to the top of the anticline, like a semi-vertical, giant coreflood, using gravity to optimize front displacement. No injectivity issues were observed: the wells were obviously fractured during the waterflood; and, given the spacing (600 m), geology and the net pay thickness (20 m), it was unlikely that any fracture could connect an injector and a producer or break through the cap rock.

The first response occurred only after one month of injection (even with a large spacing): a pure conformance effect doubled the oil production rate (from 40 to 80 m³/day for the two producers in the middle of the structure) and resulted in a 15% decrease in water-cut, which remained steady over the following 24-month period. After a 1% pore volume, high-viscosity slug was injected, it was decided to progressively decrease the polymer viscosity to 20 cP, giving a mobility ratio of 0.01. Periodically, the viscosity was increased again to higher values to maintain the conformance effect. After 24 months of injection, 170 tons of extra oil has been produced per ton of polymer injected, showing excellent performance in this reservoir. The increase in oil rate was due to this conformance effect and was followed several months later by another increase linked to an oil bank resulting from conventional polymer flooding. No polymer breakthrough has been observed in the producers up to now.

There are other examples where high-viscosity slugs were considered and injected successfully [7, 8]. This can be a very attractive method of sequencing polymer viscosity, provided the flow anomaly doesn't communicate the high-viscosity polymer directly to the producer and cause issues at the production facility.

Conclusion.
Polymer flooding can be efficient at increasing the recovery factor even in light oil reservoirs, given the presence of heterogeneities that impact overall sweep efficiency. Adapting the injection strategy by using the reservoir features can give excellent results even with difficult reservoirs where the oil saturation makes the use of EOR techniques compulsory.

Example 9.2 (Unpublished).
The field is again located in Eastern Europe, with the primary field characteristics summarized in Table 9.2.

Table 9.2

Field characteristics of Eastern European field 2

Property	Value and units
Lithology	Sandstone
Average permeability	500 mD
Injection water salinity	$257 \text{g} \text{l}^{-1}$
Divalent cations	$16 \text{g} \text{l}^{-1}$
Reservoir temperature	30 °C
Oil viscosity	70 cP

The field itself is a good candidate for polymer injection. With a recovery factor below 35%, low temperature, and relatively high oil viscosity, it is possible to substantially improve recovery by injecting a polymer. The main difficulty here is water salinity: with very high total dissolved salts (TDS) and divalent cation content, it is necessary to find a chemistry able to provide viscosity with a reasonable dosage. The fact that the temperature is low is positive: it was not necessary to develop a very specific chemistry that could resist both the salinity and temperature. A terpolymer of acrylamide, acrylic acid, and acrylamido tertiary butyl sulfonic acid (ATBS) was selected, with only 5% of sulfonated monomer, yielding 20 cP at 2000 ppm (@30 °C and 7.34s^{-1}). No dissolution problems were observed in the field with the appropriate equipment and maturation time.

Injection was performed in two wells surrounded by eight offset producers. The pilot showed incremental oil recovery of 13.7% OOIP.

Conclusion.

High salinity and divalent cation content are not showstoppers for polymer injection. Chemically resistant polymers exist that can remain stable over a long time. The main issue with high salinities is finding a polymer that can provide the required viscosity at a cost-effective concentration.

Example 9.3 – Brintnell/Pelican Lake

Canada has a lot experience in polymer flooding heavy oil reservoirs with very good results. The great thing is that all the data is publicly available: production, injection rates, well pressures, oil production, etc. Given time, it is possible to analyze specific patterns to understand the reservoir response and show the benefits of polymer injection.

The Brintnell/Pelican Lake field was among the first heavy oil fields developed with alternating horizontal injection and production wells where polymer injection proved to be successful [9]. The primary characteristics of the reservoir are given in Table 9.3.

Field characteristics of Brintnell-Pelican Lake

Table 9.3

Property	Value and units
Lithology	Sandstone
Average permeability	300–3000 mD
Injection water salinity	1–30 g l⁻¹
Divalent cations	0–1 g l⁻¹
Reservoir temperature	12–17 °C
Oil viscosity	800–80 000 cP

At first glance, the permeability, salinity, and temperature are very attractive for polymer flooding. The oil viscosity has a large range across the vast pool; however, much of the developed area would be for viscosities less than 10 000 cP. Despite this, the field has been very successful.

Several papers and presentations describe precisely the design and current results [5, 9, 10]. Since the first pilot in 1997, significant learnings have helped improve the injection strategies and field development. Currently, the entire field is operated by Canadian Natural Resources Ltd. (CNRL). In the newly drilled patterns, a polymer flood is implemented immediately after primary production (no waterflood) with tremendous results:

- Oil production with high oil cut is maintained over more than 100 months in the patterns, compared to only 40 months in the waterflood patterns.

- Injection rates are on average above those obtained for waterflood and polymer flood after waterflood, and reservoir fill-up occurs much more quickly due to less initial voidage and mitigation of water channels.

- Water/polymer breakthrough is significantly delayed.

The strategy employed is also worth looking at. In general, newly drilled patterns are produced briefly on primary production, after which the producers are converted to injectors and an infill producer is drilled with a spacing of 100 m. Polymer injection is started until the reservoir is filled and the maximum allowed pressure is reached. Once this step has been completed, the wells are put on production and the injection rates is decreased slightly to delay breakthrough.

As of 2016, current recovery factors for Pelican Lake ranged from 7% to 10% OOIP, with the original pilot location achieving 26% OOIP. The ultimate recovery factors are expected to range between 20% and 38% OOIP, depending on oil viscosity, regional permeability, and reservoir quality, with the top end of the range reserved again for the original pilot area [11].

Conclusion.

A polymer flood brings greater benefits when started right after primary production. This strategy is viable even in heavy oil reservoirs where the relative mobilities of fluids are highly unfavorable. Adapting well spacing and the injection strategy can greatly improve the efficiency of polymer injection. Other examples are the Patos-Marinza oilfield, operated by Bankers Petroleum [11, 12]; and Medicine Hat oilfield, operated by Enerplus.

9.3. Other Interesting Field Cases

9.3.1. Economic Benefits of Polymer Injection

Daqing is the largest polymer injection in the world [13], with more than 2400 polymer injectors and 150 000 tons of active polymer injected per year. An injection viscosity of 40 cP has been utilized to effectively displace an 11-cP oil and more than 12% extra oil recovery. This large polymer injection project has also demonstrated the economic benefits of injecting polymer instead of water over the long term: Yuming et al. [14] stated that the OPEX for polymer flooding are $2.83/bbl less than what is required for waterflooding. Details can be found in their paper.

9.3.2. Injection Under Fracturing Conditions

Several authors have analyzed the results of polymer injection in the field. The vast majority confirm that injection occurred under fracturing conditions: no significant pressure increase was observed when switching from water to polymer, and injectivity was always much better than expected. The idea is to keep control over fracture extension, to avoid connecting injectors and producers while maintaining reasonable injection rates. Pilot tests should be designed to test reservoir limits [15–17].

9.3.3. High-Temperature Reservoirs

An interesting recent field case is the SP flood in the Hungarian oilfield of Algyo operated by MOL. A specific surfactant formulation was developed for this hot reservoir (100 °C) and tested along with a sulfonated polymer. The entire formulation showed perfect stability over a year and was subjected to a one-month injectivity test with excellent results. Other injectivity tests have been performed in hot reservoirs in the Middle East, with publications likely to come.

Table 9.4 summarizes the published field cases after 2014. Comprehensive reviews can be found in Standnes [6] and Sheng [2]. Several field tests and projects are ongoing but have not yet been published.

Table 9.4

List of recent chemical EOR injections

Field	Company	Country	Method	References
Belayim	ENI	Egypt	P	[17]
Palogrande-Cebu	Ecopetrol	Colombia	P	[18]
Patos-Marinza	Bankers Petroleum	Albania	P	[11, 12]
Algyő	MOL	Hungary	SP	[19]
Rayoso	YPF	Argentina	P	[20]
Grimbeek	YPF	Argentina	P	[21]
West Salym	SPD	Russia	ASP	[15]
Matzen	OMV	Austria	P	[22]
ABK	Total	UAE	Single-well tracer test (SWTT)	[23]
Medicine Hat	Enerplus	Canada	P	[24]
Mangala	Cairn	India	P + ASP	[25]
XJ6	CNPC	China	P	[26]

9.4. Conclusions

Many successful pilots have been referenced in the literature. The vast majority of them have demonstrated that when the viscosity and concentration injected are high enough, assuming the wells are already fractured and the spacing allows good control over fluid flow, the injection appears successful.

In the end, the most important parameter is viscosity (or resistance factor): if all precautions are taken to ensure that the polymer enters the reservoir with little degradation (proper chemistry, clean and adapted completion, adapted equipment, non-degrading devices, and injection under controlled fracturing conditions), then the polymer solution with the target viscosity will perform as designed and will effectively displace the in situ hydrocarbons.

If we consider that viscosity/concentration (resistance factor) is the most important parameter – putting aside other rheological aspects – then it is possible to adapt the injection strategy and field design to minimize OPEX. The answers to many questions related to modeling, efficiency, reservoir response, etc. will be determined from the field response, so it's best to allocate additional capital for conducting short- or longer-term pilots or injectivity tests to gather relevant data for building a solid business case.

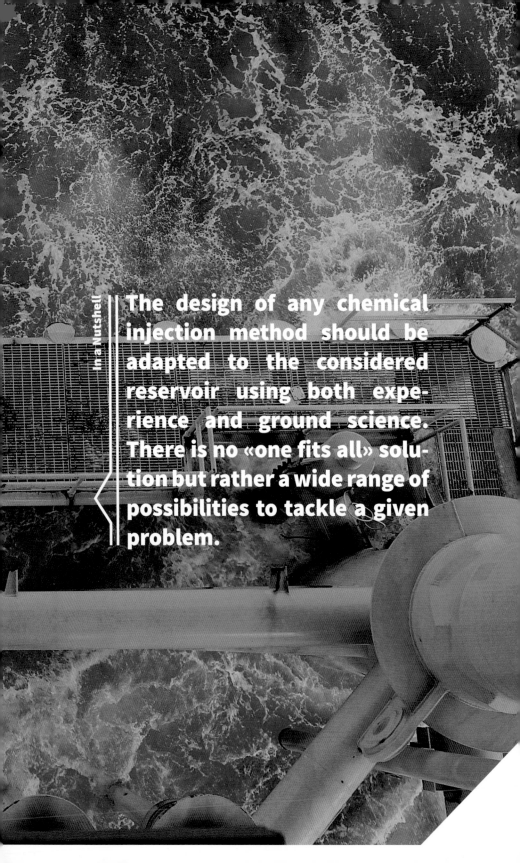

The design of any chemical injection method should be adapted to the considered reservoir using both experience and ground science. There is no «one fits all» solution but rather a wide range of possibilities to tackle a given problem.

References

[1] Seright, R.S. (2016). How much polymer should be injected during a polymer flood? Paper SPE 179543 presented at the Improved Oil Recovery Conference, Tulsa, Oklahoma, USA, 11–13 April. https://doi.org/10.2118/179543-MS.

[2] Sheng, J.J. (2011). *Modern Chemical Enhanced Oil Recovery: Theory and Practice.* Elsevier.

[3] Sheng, J. (2013). A comprehensive review of ASP flooding. Paper SPE165358 presented at the SPE Western Regional & AAPG Pacific Section Meeting, 2013 Joint Technical Conference, Monterey, California, USA, 19–25 April. https://doi.org/10.2118/165358-MS.

[4] Thomas, A. (2016). Polymer flooding. In: *Chemical Enhanced Oil Recovery (cEOR) – A Practical Overview* (ed. L. Romero-Zerón). InTech https://doi.org/10.5772/64623.

[5] Delamaide, E., Bazin, B., Rousseau, D. et al. (2014). Chemical EOR for heavy oil: the Canadian experience. Paper SPE169715 presented at the SPE EOR Conference at Oil & Gas West Asia, Muscat, Oman, 31 March – 2 April. https://doi.org/10.2118/0316-0081-JPT.

[6] Standnes, D.C. and Skjevrak, I. (2014). Literature review of implemented polymer field projects. *Journal of Petroleum Science and Engineering* 122: 761–775.

[7] Rubalcava, D. and Al-Azri, N. (2016). Results & interpretation of a high viscous polymer injection test in a south oman heavy oil field. Paper SPE 179814 presented at the EOR Conference at Oil & Gas West Asia, Muscat, Oman, 21–23 March. https://doi.org/10.2118/179814-MS.

[8] Fulin, Y., Demin, W., Wenxiang, W. et al. (2006). A pilot test of high-concentration polymer flooding to further enhance oil recovery. Paper SPE99354 presented at the SPE/DOE Symposium on Improved Oil Recovery, Tulsa, OK, USA, 22–26 April. https://doi.org/10.2118/99354-MS.

[9] Canadian Natural Resources Limited. (2016). Annual performance presentation – in situ oil sands schemes. AER Report. 9673/10147/10423/10787.

[10] Delamaide, E., Zaitoun, A., Renard, G. et al. (2013). Pelican Lake field: first successful application of polymer flooding in a heavy oil reservoir. Paper SPE 165234 presented at the SPE Enhanced Oil Recovery Conference, Kuala Lumpur, Malaysia, 2–4 July. https://doi.org/10.2118/0115-0078-JPT.

[11] Hernandez, N. (2016). Polymer flooding a multi-layered and extra-heavy oilfield. Paper WHO16-903 presented at the World Heavy Oil Congress, Calgary, Alberta, Canada, 6–9 September.

[12] Hernandez, N. (2015). Technical update on the Patos-Marinza EOR project. SPE Heavy Oil: Lifting recovery to the Next Level Workshop, Budapest, Hungary, 22–23 September.

[13] Wang, D., Han, P., Shao, Z. et al. (2008). Sweep improvement options for the Daqing oil field. *SPE Reservoir Evaluation and Engineering* 11: 18–26.

[14] Yuming, W., Yanming, P., Zhenbo, S. et al. (2013). The polymer flooding technique applied at high water cut stage in Daqing oilfield. Paper SPE164595 presented at the North Africa Conference & Exhibition, Cairo, Egypt, 15–17 April. https://doi.org/10.2118/164595-MS.

[15] Van der Heyden, F.H.J., Mikhaylenko, E., de Reus, A.J. et al. (2017). Injectivity experiences and its surveillance in the West Salym ASP pilot. Paper EAGE ThB07 presented at the 19th European Symposium on Improved Oil Recovery, Stavanger, Norway, 24–27 April.

[16] Moe Soe Let, K.P., Manichand, R.N., and Seright, R.S. (2012). Polymer flooding a 500cp oil. Paper SPE 154567 presented at the 18th SPE Improved Oil Recovery Symposium, Tulsa, Oklahoma, USA, 14–16 April. https://doi.org/10.2118/154567-MS.

[17] Spagnuolo, M., Sambiase, M., Masserano, F. et al. (2017). Polymer injection start-up in a brown field – injection performance analysis and subsurface polymer behavior evaluation. Paper EAGE Th B01 presented at the 19th European Symposium on Improved Oil Recovery, Stavanger, Norway, 24–27 April.

[18] Perez, R., Garcia Castro, R.H., Jimenez, R. et al. (2017). Mature field revitalization using polymer flooding: Palogrande Cebu field case. Paper SPE185552 presented at the SPE Latin America and Caribbean Petroleum Engineering Conference, Buenos Aires, Argentina, 18–19 May. https://doi.org/10.2118/185552-MS.

[19] Puskas, S., Vago, A., Toro, M. et al. (2017). First surfactant-polymer EOR injectivity test in the Algyo Field, Hungary. Paper EAGE ThB08 presented at the 19th European Symposium on Improved Oil Recovery, Stavanger, Norway, 24–27 April.

[20] Martino, L.A., Fernandez Righi, E., Gandi, S. et al. (2017). Surveillance and initial results of an existing polymer flood: a case history from the Rayoso formation. Paper SPE185526 presented at the SPE Latin America and Caribbean Petroleum Engineering Conference, Buenos Aires, Argentina, 18–19 May. https://doi.org/10.2118/185526-MS.

[21] Juri, J.E., Ruiz, A., Pedersen, G. et al. (2017). Grimbeek −120 cp Oil in a multilayer heterogeneous fluvial reservoir. First successful application polymer flooding at YPF. Paper EAGE Th B06 presented at the 19th European Symposium on Improved Oil Recovery, Stavanger, Norway, 24–27 April.

[22] Sieberer, M., Jamek, K., and Clemens, T. (2016). Polymer flooding economics, from pilot to field implementation at the example of the 8th reservoir, Austria. Paper SPE179063 presented at the SPE Improved Oil Recovery Conference, Tulsa, Oklahoma, USA, 11–13 April. https://doi.org/10.2118/179063-MS.

[23] Al-Amrie, O., Peltier, S., Pearce, A. et al. (2015). The first successful chemical EOR pilot in the UAE: one spot pilot in high temperature, high salinity carbonate reservoir. Paper SPE177514 presented at the Abu Dhabi International Petroleum Exhibition and Conference, Abu Dhabi, UAE, 9–12 November. https://doi.org/10.2118/177514-MS.

[24] Batonyi, A., Thorburn, L., and Molnar, S. (2016). A reservoir management case study of a polymer flood pilot in Medicine Hat glauconitic c pool. Paper SPE179555 presented at the SPE Improved Oil Recovery Conference, Tulsa, Oklahoma, USA, 11–13 April. https://doi.org/10.2118/179555-MS.

[25] Pandey, A., Koduru, N., Stanley, M. et al. (2016). Results of ASP pilot in mangala field: a success story. Paper SPE179700 presented at the SPE Improved Oil Recovery Conference, Tulsa, Oklahoma, USA, 11–13 April. https://doi.org/10.2118/179700-MS.

[26] Wu, X., Xiong, C., Liu, S. et al. (2016). Successful field test of a new polymer flooding technology for improving heavy oil reservoir recovery – case study of strongly heterogeneous and multi-layer conglomerate heavy oil reservoir XJ6. Paper SPE 179791 presented at the SPE EOR Conference at Oil and Gas West Asia, Muscat, Oman, 21–23 March. https://doi.org/10.2118/179791-MS.Table 1

Index

A

acrylamide, 93–94

acrylamide tertiary butyl
    sulfonic acid
    (ATBS), 76, 95

acrylamido-2-methylpropane
    sulfonic acid, 95

acrylic acid, 95

active alkali
    concentrations, 264

active polymer
    concentrations, 264

active surfactant
    concentrations, 264

alkali, 211–212

alkali/surfactant coreflood/
    sandpack flow tests, 42

alkali-surfactant-polymer
    (ASP), 134, 262–264

alkoxylation, 37

alkylation, 36–37

amphoteric surfactants, 35

anecdote, 7

anionic
    polyacrylamide, 143–144

anionic surfactants, 35, 36

aquifer, 81

ASP. *see* alkali-surfactant-
    polymer (ASP)

ASP flooding

economics, 44–45

field cases, 44–49

laboratory studies, 40–44

theory, 38–40

ASP formulation, 42

ASP process, 39
    alkali, 211–212

ASP reminder, 209

mixing of all products, 213

surfactants, 212–213

water softening, 209–210

associative
    polymers, 103–105

average permeability in the
    field too low
    (7mD), 272

azobisisobutyronitrile
    (AIBN), 96

B

back-produced anionic
    polyacrylamide, 143

Bahrain field, 45

bioconcentration factor
    (BCF), 143

biodegradability, 144–145

biopolymers, 92

Brintnell/Pelican
    lake, 279–281

Brookfield viscometer, 123

Essentials of Polymer Flooding Technique, First Edition. Antoine Thomas.
© 2019 John Wiley & Sons Ltd. Published 2019 by John Wiley & Sons Ltd.

POLYACRYLAMIDES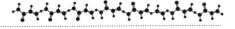

 Current recovery factor

35% *after waterflooding*

 +10% and more *with polymer flooding*

Faster oil production

Better sweep efficiency

Less water production

Essentials of Polymer Flooding

Enhanced Oil Recovery (EOR) technologies have been implemented in various fields around the world, always on a case-by-case approach. One such technique consists of injecting viscosified water into the formation to displace the oil instead of regular water. This technique is called Polymer Flooding. It has been implemented since the late 1960's with, up to now, large commercial and technical successes. This book aims to summarize the key factors associated with polymers and polymer flooding – from the selection of the type of polymer through characterization techniques, to field design and implementation – discussing the main issues to consider when deploying this technology.

polymer injection

water-cut

oil-cut

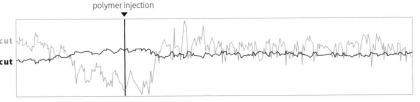

POLYACRYLAMIDES

 Current recovery factor **35%** *after waterflooding*

 +10% and more *with polymer flooding*

Faster oil production
Better sweep efficiency
Less water production

Essentials of Polymer Flooding

Enhanced Oil Recovery (EOR) technologies have been implemented in various fields around the world, always on a case-by-case approach. One such technique consists of injecting viscosified water into the formation to displace the oil instead of regular water. This technique is called Polymer Flooding. It has been implemented since the late 1960's with, up to now, large commercial and technical successes. This book aims to summarize the key factors associated with polymers and polymer flooding – from the selection of the type of polymer through characterization techniques, to field design and implementation – discussing the main issues to consider when deploying this technology.

Proven commercial history • **Simple technology** • **Quick benefits**

polymer injection

water-cut

oil-cut

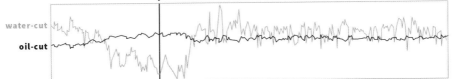